The History of
FORT LEONARD WOOD
MISSOURI

The History of
FORT LEONARD WOOD
MISSOURI

PAUL W. BASS

Acclaim Press
MORLEY, MISSOURI

Acclaim Press
— *Your Next Great Book* —

P.O. Box 238
Morley, MO 63767
(573) 472-9800
www.acclaimpress.com

Cover Design: M. Frene Melton

Copyright © 2016, Paul W. Bass
All Rights Reserved.

No part of this book shall be reproduced or transmitted in any form or by any means, electronic or mechanical, including photocopying, recording or by an information or retrieval system, except in the case of brief quotations embodied in articles and reviews, without the prior written consent of the publisher. The scanning, uploading, and distribution of this book via the Internet or via any other means without permission of the publisher is illegal and punishable by law.

ISBN-13: 978-1-942613-18-3
ISBN-10: 1-942613-18-0

Library of Congress Control Number: 2015914211

First Printing: 2016
Printed in the United States of America
10 9 8 7 6 5 4 3 2 1

This publication was produced using available information.
The publisher regrets it cannot assume responsibility for errors or omissions.

Contents

Dedication 6
Author's Notes 7
Introduction. 9
Timeline 11

Chapter One – Construction 17
Chapter Two – Namesake. 31
Chapter Three – 1940s, World War II 48
Chapter Four – 1950s, Korean War 97
Chapter Five – 1960s, Vietnam War 104
Chapter Six – 1970s, Vietnam War 115
Chapter Seven – 1980s, Cold War. 122
Chapter Eight – 1990s, War on Terrorism 164
Chapter Nine – 2000s, War on Terrorism. 182
Chapter Ten – 2010s, War on Terrorism. 187
Chapter Eleven – Future 193

Conclusion. 198
Acknowledgments. 202
Resources 203
Fort Leonard Wood Update. 212
About the Author 214
Index 216

Dedication

This book is dedicated to the many men and women who committed their lives to national defense and the freedom of those oppressed around the world. Some of these commitments were for a lifetime of military service. Some were short-term. Some were for volunteer service.

This book is about those men and women who were specifically were involved in the construction, operation and continuation of military training and technology at Fort Leonard Wood, Missouri, during the past seventy-five years. Several key political and community figures are singled out in the pages, but the book is primarily about civilians who became soldiers and civilians who built and maintained the current military base.

Author's Notes

It is amazing that many of our most significant American institutions and organizations have little recorded history currently available for the reading public. An example is Fort Leonard Wood in Central Missouri. Having retired from twenty years in higher education at a university in south Arkansas, where I developed an interest and love for writing, and having settled back home in Missouri, I began to search for a topic of research and writing of interest to me. A friend suggested writing about Fort Leonard Wood, and I discovered that there had never been an overall history of Fort Leonard Wood written.

Through fortuitous contacts I was able to meet with the recently-retired base historian, Dr. Larry Roberts. He had served the fort for twenty-two years and was very encouraging about my writing project. Besides being a wealth of information, he was instrumental in providing me contact with key base personnel. These staff members were also very supportive of my efforts and guided me to valuable research information on the base.

There were several personal concerns about writing such an important book. First was the total lack of any military experience on my part. This concern was alleviated by the encouragement of others that the absence of any military experience may provide greater objectivity in research and writing. A second concern was my lack of knowledge about military protocol. That concern was relieved by the assured promises that the base personnel would guide me through the necessary steps of military procedure.

The research on base was very enjoyable due to cooperative support from the individual base historians. Their personal accounts and suggestions were much appreciated and very helpful. Access to files and other materials provided a wealth of information. The history and current physical development of the military base is vast. It is challenging to select key portions of the history, and especially key areas of current base operations—especially from a civilian perspective.

The original intent was to have a book ready for the base's 75th anniversary in December 2015. Having begun the process in the spring of 2012, an adequate amount of time was provided for research and writing. Two years was designated for research and gathering stories of personal experiences from those who had been on base. Another year and a half was designated for writing, editing and preparing a manuscript for submission to a publisher.

My desire has been to write a book that would be of interest to the general public, and not only to military personnel. My hope is to express my personal respect and admiration for the military presence in my home state of Missouri. It is also my desire to educate the general public on the great contributions made by men and women through their experiences at Fort Leonard Wood.

Introduction

Emery Elliott temporarily left his young family and farm just outside of Houston, Missouri, early on a cold, wet winter morning in the first few days of 1941. Before leaving home, Emery got up early and completed the daily morning farm chores—milking cows, and feeding chickens, plow mules and horses. Afterward, in Houston, he picked up several neighbors and they began the one and a half hour drive. He drove his car on a hilly, gravel road along Highway 17 towards Waynesville, about forty miles north. He drove with the promise of a well-paying job as a construction worker on a newly-planned military base in central Missouri hill country.

As he approached Waynesville, he noticed the unusually heavy traffic of many other hopeful workers. He would eventually become part of 30,000 Missourians involved in completing the massive construction project, yet unnamed. Having survived the Great Depression, the new jobs offered some economic relief for families and also gave them a sense of being part of a patriotic calling.

Emery and the neighbors commuted daily for almost three months. They worked doing mainly carpenter-assistant jobs. It was an unusually wet and cold winter. They all worked with no power tools. They returned home each day after dark to do the evening farm chores. Emery eventually came down with malaria, and had to discontinue the daily trips, but he and his neighbors helped to complete the miraculous challenge of building a military training installation that now ranks among the top five of importance in the United States.

Fort Leonard Wood, Missouri. Most in the state know of its existence but have little knowledge of its significant contributions to the area, state and our nation. Few are aware that it is named after a noted United States military general who figured very prominently in the Spanish-American War and World War I. Many are not aware of its construction, closure, reactivation, growth and expansion during times of current military base consolidations and cutbacks.

For some of those who experienced basic training at Fort Leonard Wood, it became known as "Fort Lost-in-the-Woods, Misery." The choice of its location was initially questioned, but was easily rationalized. Before the creation of the interstate system, it was very difficult to get there from anywhere. The account of its basic construction, within four winter months of 1941, for preparation in training for war is remarkable. The history of the military base, among local residents and communities, is filled with interesting accounts. The development of the base into today's premiere Army Engineering, Military Police and Chemical/Biological training center is a saga of politics and practicalities. The development of young people into soldiers is also a tale of significance on the national and international scenes.

There have been interesting writings concerning a few specific periods of the base history, but regrettably there has been no account written on the history of the overall development of the base. It is hoped this book will achieve that goal along with having several other purposes. First, is to give an accurate account of the development and growth of Fort Leonard Wood; second, to give personalization to the activities on and off the base; third, to give an overview of the base's current impact on local communities, the nation and in international settings; and fourth, to instill in Missourians a sense of pride in what has happened and is happening within our state.

Timeline

1940s: German Nazi military aggression in Europe
    Base location decision, 1940
    Base construction, December 1940
    World War II declaration, December 1941
        Local USOs (Waynesville, Rolla, etc)
        300,000 soldiers trained
        Engineer Replacement Training Center
    Base closure, March 1946
    Base closure activities
        National Guard and Army Reserve annual training
        Cattle company from Oklahoma

1950s: Korean War
    Base re-activation, August 1, 1950 announcement
        30 days to prepare for 20,000 soldiers
        6th Armored Division added
    Base permanency, March 1, 1956
        United States Army Training Center- Engineer
        Dru Pippin, Committee of 50
        U. S. Army Training Center- Engineer
    Base activities
        1958, Regimental Training Areas construction
        Housing for post's permanent party personnel
            Base hospital
    Cold War training

1960s: Vietnam War training
        1966-67, almost 250,000 soldiers trained
    Cold War continued training

Cuban missile crisis, military alerts
Local community morality challenges

1970s: Vietnam War continued training
Cold War continued training
July 1973, TRADOC established
Base constructions
- Post Exchange, Commissary, Credit Union
- 1974, Truman Education Center, Bachelor Officers Quarters and Officers Club
- 1974, Interservice Training Review Organization
- 1979, Specker Barracks
- 1978, Army's first gender-integrated basic training at FLW

1980s: Cold War continued
War on Terrorism training
1982, 4th Training Brigade training engineers from 15 foreign countries
Base construction
- Duvall Maintenance Complex
- 1985, Virginia's Fort Belvoir moved to FLW, U. S. Army Engineer Center
- 1985 and 1987, FLW won Commander-in-Chief Award for Installation Excellence
- 1989, Engineer School completed move ("For the first time in nearly 50 years, all engineering training—including officers and enlisted personnel—would take place at the same location.")

1990s: War on Terrorism continued training
Invasion of Kuwait by Iraq
- Operation Desert Shield
- Operation Desert Storm
Iraq War
- 4,000+ Reserve Component soldiers mobilized, including 16 United States Army Reserve and 9 National Guard units

          1995, Base Realignment and Closure committee actions
          1999, relocate Military Police and Chemical School to FLW, $200 million expense, form the Maneuver Support Center (MANSCEN)

2000s: September 11, 2001 attack on United States by Al Queda
          War on Terrorism continued training
          Iraq War
              Operation Enduring Freedom
              Operation Iraqi Freedom
          Afghanistan War
          2002, Installation Management Agency (IMA) formed
          2008, 4th Maneuver Enhancement Brigade activated at FLW
          2009, 102nd United States Army Reserve Division re-activated at FLW
          2009, October 1, FLW designated Maneuver Support Center of Excellence (MSCoE)

2010s: War on Terrorism continued training
          Iraq War
          Afghanistan War
          December 2011, Base tornado

2020: War on Terrorism
          Future plans—Project 2020

The History of
FORT LEONARD WOOD
MISSOURI

Chapter One

CONSTRUCTION

Introduction

On December 3, 1940, in a small clearing on a wooded ridge in the Missouri Ozarks, a sharp, damp wind greeted the small group of military personnel and construction workers for an inauspicious groundbreaking event. What was described as the "first shovelful of muddy, frostbitten earth" was turned over, history was changed for the state of Missouri, the United States and the world. Events leading up to and following this groundbreaking were monumental.

It appeared all too certain that the isolationist policy of the United States after World War I was about to be challenged, as the world prepared for another war. That First World War provided several important lessons to the United States. First, was a total lack of adequate training and technology needed to provide national involvement for a war in foreign lands. Second, was the horrible price such an effort cost in terms of human lives and material resources. Third, was the need for a united national commitment to involvement in such a war.

By early 1940, most of Europe had fallen to the overwhelming force of the German Nazi war machine under the leadership of Adolph Hitler. France and Great Britain were being seriously threatened with the prospect of a German invasion. The British decision for heroic opposition to Hitler caused the United States to contribute indirectly to the war effort through a Lend-Lease program of arms and equipment to the British. To President Franklin Roosevelt it appeared that the United States must evaluate its preparedness for the inevitable direct involvement in a second world war.

The U.S. Army in 1940 was ranked eighteenth in the world, behind Germany, France, Britain, Russia, Italy, Japan, China, Belgium, the Netherlands, Portugal, Spain, Sweden, and even Switzerland. Germany had years of required military training establishing a trained military of 6.8 million, about ten percent of the population. By contrast, the total U.S. military forces on active duty was 504,000, about 0.5 percent of the population. When Germany launched her European offensive in May 1939, she was supported by 136 divisions; the United States could barely muster five fully equipped divisions. In addition the United States, still reacting from World War I, had backed away from hardly any arms and munitions productions.

Any nation preparing for war must consider resources—human and material. In terms of human resources, the troop strength of the United States in the fall of 1939 was only 190,000. The armaments and war materials were still vintage 1920 variety. With the new Chief of Staff, George C. Marshall, a new era of military leadership began for the United States Army. An ineffective and outdated military organization was to be changed under Marshall's leadership. New training procedures and manuals were developed. There was a pressing need for a national military draft, as requested by Marshall, to enlist five times the number of active soldiers. Roosevelt's major hurdle in initiating war preparations was his secretary of war, Harry Woodring, who was a strong isolationist. Woodring was replaced by Henry L. Stimson, an admirer of George Marshall. In June 1940, France fell to German occupation. Congress and the president then acted quickly to approve the Burke-Wadsworth bill, establishing a national draft in the Selective Service Act of September 1940, in spite of the looming fall elections. Later, when the act was to extend the twelve months of required active service with an additional eighteen months, passage in the House of Representatives, after extensive lobbying, was by the single vote margin of 203-202.

With the sudden increase of troops needing to be trained, many logistical problems were created. There were an inadequate number of training facilities, and most of these were on the east and west coasts and the Old South. Only one Midwestern training facility was under construction near Little Rock, Arkansas. It was to be known as Camp Robinson, named after Joseph T. Robinson, former Arkansas representative, governor and senator. There was an immediate need for another, more northerly, military facility to be located in the Midwest.

To provide adequate training facilities for the expected 1.4 million conscripted soldiers, forty-six new army bases had to be located and constructed. The job described was defined as monumental: "...land had to be cleared, hills leveled, valleys filled, trees uprooted, roads

Construction

surfaced, and drainage systems installed before construction of barracks, laundries, officers' quarters and rifle ranges… The building of new camps required 400,000 men, 908,000 gallons of paint, 3,500 carloads of nails, and 10 million square feet of wallboard." Work would not begin until after passage of the Selective Service Act.

Site Decision

As might be expected in making such a decision for the new site, many considerations were taken into account. The obvious availability of natural resources, social impact, training space and political impact were all-important facets of the decision. General Leonard Wood's warning, decades earlier, that the immediate employment of mass U.S. troops, inadequately trained, to Europe would result in unnecessarily large numbers of injuries and deaths. Such warnings were still true thirty-five years later. A newspaper article reported the headline, "One Commander Says 200,000 men would be lost in a hurry if Americans were thrown against Germans today—Answer is more training."

An earlier decision by the War Department had already decided to establish a military training center in southern Iowa, at Leon. This would accommodate the training for the Seventh Corps and the Sixth Infantry Division of Minnesota. Local communities in Iowa and the state politicians welcomed the decision and anticipated all of the future financial benefits. When the military engineers investigated the proposed site further, it was discovered that the local water table had dropped considerably—sixty feet since 1918. The financial feasibility of the location was now in serious question. There was also the question of whether the proposed Iowa camp had adequate acreage for the varieties of training envisioned.

As word spread of the problems with an Iowa location, Missouri cities sprang into public relation campaigns to secure the military training site. Kirksville and Springfield chambers of commerce contacted the War Department extolling the wisdom of relocating in one of their cities. It is certain that Missouri politicians played a role in investigating a Missouri site. One prominent U.S. Senator, Harry S Truman, having left limited correspondence on record, was becoming a major player in national military issues. When pressed by the Missouri cities for his assistance in this matter, Senator Truman replied:

> …these Training Centers are placed by the National Defense Council and the Army Engineers and it

is a most difficult matter to get them located in any specific place in the State. I will be glad to do whatever I can however.

Iowa was certainly not happy with the prospects of losing the military facility. The Springfield, Missouri Chamber of Commerce would write to Senator Truman:

> We have word today that Senator Clyde Herring, of Iowa, is going to make a vigorous protest to the President because the SEVENTH CORPS AREA TRAINING CENTER is to be located in South Missouri rather than in Iowa as originally planned.
> We urge you to do everything possible at once to preclude this possibility.

A reply from Senator Truman's office was given to the Springfield Chamber of Commerce:

> From the information that we have been able to gather here, I don't think there is any question but that this Training Center will be located in Missouri.
> We have been watching the situation here and doing everything possible to see that it is located in Missouri.

Department of Agriculture and forest service men did advance scouting for the Army. Upon their findings, Army engineers and the architect-engineer firm of Alvord, Burdick, and Howson of Chicago, Illinois investigated land just south of Waynesville in central Missouri. They found several strong advantages for consideration of the area. Water supplies were in abundance with the Roubidoux and Big Piney Rivers. Numerous large, fresh water springs were located in the immediate area. The local terrain offered a diversity of training and engineering opportunities. Another big plus was the already established Mark Twain National Forest. The surrounding area was very sparsely populated and would offer minimal unfavorable publicity due to displacements for the needed land acquisition.

Significant negatives of the location were transportation and

Construction

communication. Although the area was strategically located between St. Louis, Kansas City and Springfield, the roads in 1940 were a problem. Route 66 was about the only concrete, two-lane road in the area. Most of the others were gravel or dirt. The nearest railroad line was 20 miles east in the small community of Newburg. The nearest major telephone base was in Rolla, about 30 miles east. But similar negatives were also associated with the Iowa site. The Missouri site seemed to be preferable.

A decision was made by the War Department in October 1940, that the new military base would be constructed just south of Waynesville, Missouri, in the Gasconade Unit of the Mark Twain National Forest. Initially, it was to be called the Seventh Corps Regular Army Area Training Center. The purpose was to create "a divisional training center and would also be the site of an Engineer Replacement Training Center (ERTC)." The announcement took local residents by complete surprise.

Land Acquisition

With the site selection now made, action would begin immediately to prepare for construction of the military post in this challenging, rugged area. In November 1940, Army surveyors would begin the task of planning for the site preparation. The Iowa camp was to contain 23,000 men at an estimated cost of $8 million. The Missouri facility could now be expanded to 35,000 men with an estimated cost of $35 million. To understand the environment in which surveyors, construction workers and Army engineers would begin this time-sensitive task, it is important to know about the land, the people and the communities. An extensive historical examination of these areas is found in Steven Smith's very interesting 2003 book, *Made in the Timber: A Settlement History of the Fort Leonard Wood Region*.

In the early 1800s, settlers made their way to an area of harsh landscape and poor prospect for cultivation. Land around the rivers and streams were settled first. Missouri was approved as the 24th state in 1821. In the winter of 1838, the area was a stopping place for the Cherokees on the Trail of Tears. The Homestead Act of 1841 provided incentive for additional, but limited, settlements. The small village of Waynesville (named for Revolutionary War hero "Mad Anthony" Wayne) was established as the county seat in 1843. By 1860, the area of Pulaski County (named after the Revolutionary War

Polish General, Count Casimar Pulaski) averaged about seven people per acre.

During the Civil War, soldiers described the area as "a miserable apology for a village" and "a sorry looking place." The population lived a subsistence lifestyle suffering from the effects of weather and wilderness. The Civil War also saw battles and residents favoring both the Union and the Confederacy, often in the same families. The area was under the predominant control of Union forces. Some dropouts from the Civil War found sanctuary among the hills and woods of the area, eventually settling there. A local industry developed around the timber available for railroad ties and construction lumber.

World War I found a limited number of recruits from the area joining the 156,000 Missourians serving their country in the war effort. Local supplies of wood and minerals aided the war effort. The effect of the Great Depression on the area was comparatively minimal, but lasted longer than in most other areas. The Roosevelt programs such as the Civilian Conservation Corps provided some temporary economic assistance and some additional structures in the area. A number of camps were established within the Mark Twain National Forest. With the announcement of the construction of the new military installation, the population had mixed reactions.

War Department planners estimated that 65,000 acres were initially needed for establishment of the base. The Mark Twain National Forest land available, being owned by the government, amounted to 12,000 acres. Another 53,000 acres would be needed in the surrounding area, which would come from local landowners.

Securing of the additional land needed would not be an easy process for the local population. The takeover was rapid, with much of the land being purchased by Christmas, 1940. Some landowners began a crash course in property improvement, hoping to ask for a higher price. Some families saw the sale of their land as a patriotic duty in war preparation. A few families were just glad to get rid of their property. There were over 300 families in the area to be relocated from the 53,000 acres purchased by the government. Of the 300 families, only about 200 were actual landowners. They were given a deadline of moving by March 1, 1941. Even though the population was sparse, several significant and meaningful communities had sprung up in the area. Among these were Bloodland, Cookville, Evening Shade, Palace, Tribune and Wharton. There was great resistance in these small communities that already had a few business establishments, school houses, churches and cemeteries in addition to homes.

The struggles experienced by these families is illustrated in the reports of older couples to Albert Mussan, agricultural economist. One couple stated, "We hate to be put out of our home. We are getting old. We can't stand changing around too much." Others saw a degree of patriotic duty in moving from their land, "Guess though, we should do our part in preparing for war and hope we'll never have to fight." Others were more pragmatic in their understanding, "The defense program needs land so why shouldn't it be us to sell and move out as well as any one else, I've been waiting to sell my no-account land and that the chance has come I sure don't aim to kick up no rumpus and act like I don't want to leave." Understandably change, fear of the unknown and compulsory eviction caused personal and communal concerns.

By July 1941, it was reported that fifty-nine families had still not moved. The remaining families sought intervention from government agencies, and even appealed to Senator Truman for some reprieve. Arbitration and compromises were made to most of the other families. One example was the cemetery in Bloodland, the largest of these communities. The residents were assured that the base would provide upkeep for the cemetery and access to the cemetery would be allowed on each Memorial Day "to honor the dead and renew old friendships and memories." This custom was continued for many years until many of the original residents were deceased. During that time some original residents used the opportunity to complain about the cemetery upkeep—justified or not. The loss of these communities would continue to be a source of bitterness for some of the residents. About one half of the families relocated in Pulaski County. Most of the other families resettled in other Missouri locations. But the financial benefits to the area far outweighed the dissatisfaction of the resettled families.

Initial Preparations for Construction

With the land—65,000 acres—finally secured, the task of designing and preparing for construction was well underway. The requirements seemed insurmountable. Access to the area was needed. Roads to the area were limited and roads within the area were in very poor condition. The nearest rail system was 20 miles east in Newburg. There were poor means of communication in 1940. All utilities were needed. Housing and staff buildings were needed for 32,000 recruits and 1,200 officers. Training sites were required, with safe location of artillery ranges. All of this was to be done in a secure environment and to be completed as soon as possible.

There were numerous obstacles to be faced as construction began. The terrain, while ideal for military training, was terrible for the construction process. Even though a small portion of the area was cleared for homesteading, most of the area remained as forest and wilderness. The weather proved to be a formidable challenge. The winter of 1940 was one of the wettest and coldest in many years. Mud and frozen earth would hamper the construction progress for man and machine. The availability of required labor force was also non-existent in the immediate, local area. Even housing for such a large labor force was unavailable. There was also the approval needed of federal budgets as well as the cooperative efforts of state and local governmental agencies. All of these challenges were complicated by the urgency of completing base construction as needed by the goal of three months—February 15, 1941.

After all of the initial planning and surveying were completed, the first process was securing reputable construction firms. The initial, main responsibility for the construction phase was under the control of the Construction Division of the Quartermaster's General's Office. In December 1941, the remaining construction was placed under the responsibility of the Army Corps of Engineers. The scope and timing of the project required more than one construction firm. Four construction firms were selected to begin the process—W. A. Kilnger and Sons of Sioux City, Iowa; Arthur H. Neumann Brothers, Inc., Des Moines, Iowa; Western Contracting Corporation; and C. F. Lytle Company, also of Sioux City, Iowa. After that, subcontractors were hired to begin the process. The first job was clearing the land. This would prove to be a challenging undertaking.

James Berry was a bulldozer operator in Springfield, Missouri. In the winter of 1940, he left his new bride and took his heavy equipment to the Waynesville area. According to his wife, Alabama, he "was in charge of the first clearing of the land for the fort." He was very creative in reversing the fans of the dozers and providing curtains for them to keep the operators warm. The job of removing mostly post oak trees in the mud and cold was very difficult. Just keeping the machinery running was difficult. His wife joined him later near the fort when he was able to buy a small camper that he underpinned for warmth. Their trailer was located near the White Ripple Club, an establishment, to her, of very questionable repute. James worked for many months before being hired to work at another military base under construction.

Construction Process

With the land being cleared, the task of securing laborers was underway. The area in the winter of 1941, was still recovering from the Great Depression. Jobs were scarce and the prospect for a paycheck was welcome news for many families in the area. Many workers came from within a one hundred-mile radius of the base. One week after the ground-breaking on December 3, a steady stream of workers, equipment and supplies began arriving in the area overnight. Over 7,000 vehicles a day inched along the two-lane Route 66. In early fall, 1940, Waynesville had a population of about 462. Within a few months, its population increased ten-fold to over 4,000. Housing was obviously the greatest immediate need. There were a few large boarding houses in Rolla and Newburg.

The president of the Rolla Chamber of Commerce wrote to Senator Truman in February of 1941 about the problems the area faced with construction workers and their families taking over housing designed for the college students. The desire for new housing construction was hampered by the requirement of sewage expansion. That same month an officer of the Fox Amusement Corporation wrote to Senator Truman complaining that the plans to build a new theater in Waynesville was stopped because of the need for expanded water and sewage service. The locals saw the financial opportunity and creatively sought to meet the need. Every little storage building, garage, basement, chicken coop, shanty, woodshed and spare bedroom was converted to bedding space. Some facilities allowed for a system of triple bunk beds. The workers would also be divided into three, eight-hour shifts working day and night. In many places men would buy the bed space and a meal for the eight-hour shift at $2.50 a week. Some workers paid the 80 cent bus fair to Rolla and stayed in a comparatively nicer flop house. Camps—shanties and tents—sprang up overnight by the main and access roads near the streams and rivers. Many slept in cars and trucks until any space was available. The peak of the labor force by early spring, 1941 amounted to 30,775 workers living within a fifty-mile radius of the base. This was the largest work force in the nation at that time. In addition to the laborers, there were 500 engineers and 1,006 office personnel. The peak payroll for one week, paid out on March 22, 1941, was the largest regular single payroll in the United States—$1,342,418.79. As the first of the base barracks was completed, by January 3, 1941, fortunate workers were able to claim beds. The rooms were warm and 36-cent meals were available on base.

One of the workers was Virgil LaJuene. He was an electrician and left Springfield to work on the base in barrack construction with the Barrett/Hipp construction company. He and his family, a wife and two children, camped along the Gasconade River in a small camper. After the completion of their 500th housing facility on base, the family left Fort Leonard Wood, coincidentally on December 7, 1941.

The workers were generally paid 75 cents an hour—a good wage in the winter of 1940. Another problem soon developed for the workers—cashing their government paychecks. The usual paychecks totaled more than $200,000 each week. The smaller communities and even the larger communities were unable to meet the demand for cashing such a volume of checks. The government eventually had to provide armored cars with sufficient cash for the workers.

As might be expected, government purchase of land and the crushing need for housing created a real estate boom for the surrounding area. Prices soared overnight with speculators, both local and statewide, seeing the potential profits. The rapid growth of diners and places of entertainment, especially along the main highways in Pulaski County and near base entrances was also booming. Prices for rental spaces were excessive—often displacing the poorer locals. Some of the "entertainment" businesses were scenes of illegal and immoral activities. Gambling and prostitution were commonplace, with workers looking for ways to spend their allowances. This was not unique to the base near Waynesville. Sadly, such was the case around every military base in the country.

Senator Truman's office was inundated from late 1940 through the spring of 1941 with requests for assistance in obtaining concessions on the new military base. There were letters and telegrams from all variety of business interests: Beer, barber shops, food items, dairy products, post exchange, potatoes, laundry, banking, garbage disposal, construction work and even funeral services. Senator Truman explained his inability to assist with such matters and referred these

Construction

queries to the base commander yet to be announced. There were also inquiries from real estate companies and private citizens about the long range plans for the military base. Senator Truman replied in late 1940 and through the middle of 1941:

> I have no way of knowing whether Fort Leonard Wood will be a permanent post or not. I am of the opinion it will be due to the fact it is centrally located so far as the Seventh Corps Area is concerned, and far enough South for all to use. Anyway, it will be in operation for two or three years, and conditions will take care of themselves if the emergency should suddenly stop, but I believe it will be used as a permanent post.

With the arrival of workers; construction materials, equipment, and supplies were also being gathered. This further congested the already crowded highways with steep grades. There was a need for a railroad to increase the amount of construction materials and eventual arrival of recruits to the base. The nearest line was 2.5 miles west of Newburg, twenty miles from the base. The first grading on the new railroad began on December 5, 1940, just days after the groundbreaking. The project was a major engineering challenge—clearing thick woods, crossing rivers and streams, going up steep hills, crossing deep valleys and harsh winter weather conditions. A crew of 2,800 workers completed the two-year project in five and half months at a cost of $2.5 million dollars. It was reported that

> More than 1.6 million cubic yards of dirt and rock were moved to complete the 68 cuts and 68 fills.... The steepest grade was 2.26 percent and the longest grade was 6.17 miles. At one place the Ozarks had been slashed 46 feet for the deepest cut, while hills were leveled in another sector by a fill 60 feet high. Of the 19.85 miles of rail, the longest stretch was only 2,700 feet, while 70 curves were necessary because of the terrain.

Where the railway entered the base, covered warehouses were constructed to store materials and supplies. The first train to make the run into the base was in May 1941. In the first year, the post began receiving 1,500 tons of supplies each day. The trains also provided the means for excursions for 700-1,000 soldiers each week.

When contractors, subcontractors and suppliers are involved in any government contract arrangement, careful oversight must be maintained. Such was the case with the Fort Leonard Wood construction project. Senator Truman's office had received word about some corruptive practices on the base. There were over one hundred letters complaining of the "waste of manpower and material" at Fort Leonard Wood and Camp Crowder in Missouri. Instead of just writing letters about the questionable activities, Truman took a hands-on approach. He determined if these activities were happening in Missouri, they must also be happening at other base constructions in the country. During a four-day Senate recess in the spring of 1941, he drove his own car to several bases under construction near the nation's capitol. He was largely unknown nationally and was perceived on these bases as a potential contractor. His investigation supported the accusation of corruption. Truman returned to the Senate and reported his findings to Senator Bob Reynolds, chairman of the Military Affairs Committee. Truman was a member of that committee. Senator Reynolds saw no need for any further investigation. Unsatisfied, Truman reported his findings officially on the floor of the Senate. After his report, the Senate approved a special committee to investigate the findings and made Truman the chairman. He was given a budget of $15,000 for committee expenses. Over the next five years the committee was given an additional half-million dollars to investigate the practices at Fort Leonard Wood and other military installations under construction. It has been estimated that Truman and his committee saved the United States eight to ten billion dollars in potentially wasted spending.

The official name of the military base was changed from the Seventh Corps Area Training Center to Fort Leonard Wood by the War Department General Order Number 1, issued on January 8, 1941. The designation of the base as a "fort" instead of a "camp" was to indicate its permanency. That would later be a source of tension for the Army and the local communities.

The unrealistic goal of completing Fort Leonard Wood for training by February 15, 1941, was changed to May 25, 1941. Weather delays, transportation congestion, and coordination of materials and supplies contributed to the strenuous physical demands. With state cooperation, some of Route 66 was improved with the addition of four-lane sections. In late spring, 1941, the weather finally improved speeding up construction progress.

Construction Accomplishments

By the end of April 1941, within six months of the groundbreaking, a construction miracle had been performed at Fort Leonard Wood. A total of 1,600 buildings were constructed using 75 million board feet of lumber. Among those buildings were 600 barracks, 205 mess halls, five theaters, nine hospitals, 24 filling stations and 146 day rooms. At the peak of construction, one building was completed every forty-five minutes. Other projects completed were 381 fire plugs, 60 miles of water mains, 52 miles of sewage lines, 900 manholes and 58 miles of roads. Permanent power connections were completed in June 1941, with more than 2,500 utility poles and 2 million feet of wire.

The local communities also felt a construction impact. Rolla, Waynesville and Newburg saw immediate increase in housing and office construction. Rolla became the center for applicants wanting to secure the staff support job openings at the base. In Rolla, long distance phone calls increased from 250 calls to 850 calls a day. Six new long distance wires were added to Newburg, ten to St. Louis, ten to Springfield. Seventy-five new phones were installed in Newburg.

Conclusion

As the nation moved toward direct involvement against military aggression in Europe, an honest evaluation of preparedness loomed large. The draft would provide the needed human resources. The newly-constructed military bases would provide the places for training. The natural resources of the nation would begin to radically change to provide the military resources and updated technology to fight in a contemporary war.

Although the nation was still divided over involvement in matters of Europe, with isolationists sounding the alarm, the nation would become united and resolved to answer the call to war. That simple groundbreaking in the Ozark hills of Missouri on December 3, 1940, led to a monumental construction effort completed six months later. Then seven months later, after December 7, 1941, the nation would rally around the president's and Congress's official declaration of war. The earlier measures to develop military bases such as Fort Leonard Wood would prove to provide a great advantage in immediate mobilization of a nation prepared for another international war.

Chapter Two
NAMESAKE

Introduction

A new military base was under construction. Conscription for military service was underway. Facilities for housing and training were being completed. The name of the new military base was yet to be determined. The naming of any military post is significant to illustrate its mission and identity. There are political and military considerations involved. There are usually different opinions of the best and most appropriate name. Such was the case with this newly-constructed base in Missouri.

Two names were under serious consideration. Missouri-born General John J. Pershing was the commander of the American Expeditionary Forces in Europe during World War I. He achieved international and national acclaim for his distinguished service. There would be significant local support, especially in Rolla, for the Missouri post to be named after Pershing. General Leonard Wood, whose legacy had sadly been lost to history, made significant contributions to the United States military. Although he provided the essential foundation of turning great numbers of young people into soldiers, very little is commonly known about him even today.

To better understand each man's significant military contributions and the naming decision that was made, an understanding of their lives and experiences is necessary. Each began their careers apart from consideration of any military service. Both were contemporary, American military heroes. Each married into prominent and influential national political families. Each suffered from repeated bouts of physical ailments. Each rapidly advanced to high military rank over hundreds of their superiors because of the common practice of political influence in Washington, D. C. Accordingly, each developed an influential list of political friends and enemies. However, each also developed different personal habits and military insights that distinguished their service to their country.

John J. Pershing

John Joseph Pershing was born in Laclede, Missouri, on September 13, 1860. His early education in a special school for intelligent youth provided the necessary foundation to begin a teaching career. In 1878, he taught at a school for African Americans in Prairie Mound, Missouri. In the summer, he continued his education at the State Normal School at Kirksville, Missouri. In 1882, he applied for entry at West Point because of the promise to provide an elite, college-level education. His acceptance led to an interest in the military and an opportunity for career advancement.

After graduation from West Point in 1886, Pershing was assigned to the 6th Calvary at Fort Bayard, New Mexico. He took part in several campaigns against the rebellious native American Apaches and Sioux. Pershing was cited for his bravery and meritorious service. In 1887, he was transferred to Fort Stanton, where he excelled in rifle marksmanship. In 1890, he was transferred to South Dakota to put down the last of the Sioux uprisings. In 1891, Pershing accepted a position at the University of Nebraska as Professor of Military Science and Tactics. Dismayed at the apathy shown by students, faculty and community, Pershing was fighting against the "accepted recipe for army-making" by then Secretary of State William Jennings Bryan that "a million men (would) spring to arms overnight." Pershing fought against the popular tide of thought and successfully developed an award-winning military drill company. He also completed studies for a law degree, graduating in 1893.

In 1895, Pershing was assigned to the 10th Cavalry, one of the first "Buffalo Soldier" regiments, at Fort Assiniboine in Montana. There he again led troops rounding up renegade Creek Indians and transporting them to Canada. He became a strong spokesman for African American troops. In 1897, he was assigned to West Point as an assistant instructor in Tactics. Because of his assumed strict disciplinary actions, the cadets began referring to Pershing as "Black Jack." In 1898, at the request of his superior officers, Pershing was assigned to duty in Washington, D.C. as director of the newly-developed Division of Customs and Insular Affairs. After recovering from a bout of malaria, he began his duties to provide military government for the new insular positions of the United States: Cuba, Puerto Rico, the Philippines and Guam. In 1899, he was ordered to Manila in the Philippines as part of the 8th Army Corps. For two years Pershing participated in rounding up the insurrecting native groups. His distinguished service and actions caught the attention of his superiors, even the Secretary of War, Elihu Root. In 1901, Pershing was eventually transferred to the 15th Cavalry unit in the Philippines. There he became knowledgeable of the area and people. He learned to speak the Moro language. This aided his assignment as intelligence officer and proved successful in assisting him with responding to attacks on Camp Vicars where he was stationed. In 1903, Captain Pershing was ordered back to the United States. In Washington, D.C. Pershing met and eventually married Miss Helen Frances Warren, daughter of a Wisconsin Senator, Francis E. Warren, who served as chairman of the Senate Military Affairs Committee.

In 1904, Pershing was assigned to duty in Oklahoma City as assistant chief of staff, Southwest Division. In 1905, at the request of President Theodore Roosevelt, Pershing was a military attaché to Tokyo and served as an observer of the Russo-Japanese War.

In 1906, Pershing was promoted by President Roosevelt from captain to the rank of brigadier general, over the heads of 862 more senior officers. This controversial promotion created political and military enemies, especially in light of his relationship with the Wisconsin Senator. That same year, Pershing was reassigned to the Philippines at Fort McKinley near Manila. Pershing was instrumental in establishing peace with the continually rebellious Moros. In 1913, he was assigned to Fort Bliss, Texas, commanding the 8th Brigade. During this time Pershing was dispatched to Mexico to challenge the Mexican revolutionaries led by Pancho Villa. In 1915, as Pershing was preparing to bring his family to the Southwest, he received word that fire had destroyed his family home killing his wife and three of his four children. Only his son, Warren, remained. After the funerals in Cheyenne, Wyoming, Pershing returned to Fort Bliss to lead the "Punitive Expedition" to capture Pancho Villa. Although the primary objective was not accomplished, Pershing's command and tactics attracted the interest of a young first lieutenant, George S. Patton.

In 1917, the United States declared war on Germany and Pershing was ordered back to Washington, D.C. He was eventually assigned the duty of commander of the American Expeditionary Forces. While the title of the command was impressive, the reality was very different. The United States at that time had only a regular Army of 25,000 men under command. There were no reserves as such except for the sparsely-numbered, trained men from officer's training camps of the Plattsburg Movement (established by General Leonard Wood). It was estimated that 500,000 men were needed to complete the necessary military organization. Pershing's responsibility was not only to organize the United States Regular Army under such expansion, but to persuade the British and French forces that the United States was not providing men to fill their military commands, but to establish a separate United States Army in Europe. It took months of recruitment and training in the United States to send men to Europe. Pershing took the responsibility for special training and involvement of the U.S. troops in European actions. The integration of the U.S. troops in the Allied war efforts brought heavy casualties, but eventual victory over a two-year period. With the Armistice signed on November 11, 1918, Pershing had commanded troop strength of 1.8 million men. He was a returning war hero who wisely left war reconstruction activities and policies to the politicians.

Upon his return in 1919, Congress authorized the new rank of General of the Armies and promoted Pershing as the only general with that rank. In 1920, some Republicans wanted Pershing as their candidate for president. He refused to campaign but offered to serve if nominated. Because of his close association with Democratic president Woodrow Wilson's policies, he did not gain strong support. In 1921 he was appointed chief of staff of the United States Army. In 1924, he retired from military service and became a private citizen. In 1932, he won the Pulitzer Prize for his memoir, *My Experiences in the World War*. He continued as a strong advocate for military preparedness for the United States. As he followed the deteriorating actions in Europe, he strongly supported the aiding of Britain in preparation for the second world war. Pershing was able to see the United States' involvement in World War II and the eventual Allied victory in 1945.

John Joseph Pershing continued to be recognized and honored by a grateful nation. He died at Walter Reed Hospital in Washington, D.C. on July 15, 1948. There were many tributes of honor paid to him in the United States and around the world. It would be a strong argument to favor naming the new military base in Missouri after a native son.

Leonard Wood

Unlike Pershing's Midwest beginnings, Leonard Wood was an Easterner. He was born on October 9, 1860 in Winchester, New Hampshire. Born into a home where his father practiced medicine, Leonard was strongly encouraged in the direction of that vocation. He was also raised as an outdoor person, and developed great physical strength and stamina. Upon his father's death in 1880, Leonard entered Harvard Medical School. While in training as a surgeon, he accepted a junior intern position at Boston City Hospital. Strict policy required any intern performing any surgical procedures to do so in the presence of a regular surgeon. Leonard found this policy unacceptable, and performed minor surgeries on his own. After repeated warnings, and after performing a skin graft operation, he was dismissed from the hospital in 1884. Undeterred, he "bought out a friend's practice in the tenement section of Boston." While attending to mostly charity medicine, he received a medical certificate of questionable accreditation. The personal qualities of insubordination and disregard for policy would become hallmarks throughout his career.

In 1885, Wood decided to enter the Army Medical Department. Because of his lack of a qualified medical degree, Wood did not receive the full military benefits of an Army physician. He was assigned to Fort Whipple, headquarters of the Department of Arizona. He was then assigned to Fort Huachuca, headquarters of the 4th Cavalry, near Indian country. During the long ride to the new fort, Wood explored the land and acclimated to his environs quickly. The Apaches were in rebellion and established small raiding parties causing destruction and death in Arizona and northern Mexico. The cavalry units made what were called "Indian expeditions" to try and stop the raiding parties. While most officers and soldiers tried to avoid these arduous treks, Wood quickly volunteered for field duty. He became known for his physical prowess as well as his medical practice.

In 1886, a major year-long campaign began against the Apache chief, Geronimo, and his band of notorious renegades. Defying orders to be confined to a reservation, the group began a gruesome practice of killing and property destruction. While the U.S. Cavalry was in pursuit, Mexican nationals also organized to hunt down the renegades. Wood again volunteered for field action. He gained recognition of his superior officers for his bravery and endurance, even in light of physical injury. Because of the loss of leadership during a lengthy expedition, Wood actually assumed leadership command of his unit. Although their efforts to capture Geronimo and his band were unsuccessful, through these efforts along with the threat of the more dangerous Mexican nationals, Geronimo was eventually persuaded to surrender. In 1889, Wood was singled out to receive the Medal of Honor. There was strong resentment from the ranks of the Cavalry units, "when so many line officers might have been rewarded, the government chose a contract surgeon in the medical department." The resentment and criticism would follow Wood throughout his career and created a number of future, influential opponents.

The experiences of Wood in the Southwest provided the opportunity for military advancement. Wood could now claim "a successful Indian campaign and an influential sponsor[s]." In 1888, at the insistence of one sponsor, General Nelson A. Miles, Wood was transferred to the Division of the Pacific at the Presidio in San Francisco, California. The barrenness and isolation of the Arizona fort were replaced with the activities and social life on the California west coast. Wood participated in the new experiences and encountered Louise Condit-Smith, vacationing with her guardian uncle, United States Supreme Court Justice Stephen Field. After a brief courtship and personal inquiries from Justice Field, Wood and Louise were married on November 18, 1890, in the home of Justice Field in Washington, D.C.

During this time, with his newly-established family, Wood continued to be involved in military activities, such as marches and campouts. He also expanded his physical pursuits of football and boxing. He continued his desire to be transferred to the regular army. The comparatively enjoyable life on the west coast was ended when in 1893, Wood's two-year deployment ended. He was transferred to Fort McPherson, Georgia. While there, Wood participated in the usual military activities and expanded his interest in football. He organized a team at the nearby Georgia Tech in Atlanta. Wood eventually became its coach and star player. As his two-year deployment was nearing an end, Wood sought assignment to fill the vacancy of assistant attending surgeon in Washington, D.C. There was a vigorous campaign for the position in opposition of the desires of the Surgeon General. Wood personally campaigned for the position going around and above the Surgeon General. The Secretary of War intervened, and in 1895 Wood received the desired assignment. Again, the insubordinate activities of

Wood created resentment and additional enemies. Wood was seen as developing two distinct personalities—great energy and activity as well as self-serving ambition. His location in the nation's capitol provided the opportunity for long desired military advancement. Wood was learning how to use the political process to his advantage.

In his official capacity, Wood encountered prominent political figures, including cabinet members and Presidents Cleveland and McKinley. When McKinley became president, Wood was invited to be the personal physician to the invalid first lady. This daily access to the White House and its activities, provided opportunities for political as well as social contacts. At one of the evening White House parties, Wood met Assistant Secretary of the Navy Theodore Roosevelt. Their mutual interest in physical activities, outdoor adventures and western experiences created a strong bond and mutual friendship that lasted for the rest of their lives. Involvement in Washington politics also led Wood to an acceptance of the prevailing doctrine of expansionism, especially popular with Roosevelt and new friend, Congressman Henry Cabot Lodge.

The rise of rebellion in Cuba over domination by Spain created an atmosphere of interest in eliminating European influence close to the shores of the United States. The suspicious sinking of the battleship *Maine* on February 15, 1898, increased the possibility of war with Spain. Wood and Roosevelt tried in vain to seek commissions in the military organizations sure to be involved in such a conflict. In April of that year, open hostilities began with Spain. Congress authorized the creation of three volunteer units to be commanded by federally-appointed officers. Wood and Roosevelt used their considerable influence with President McKinley, cabinet members and congressmen (especially Lodge) to gain authorization to form the first United States Volunteer Cavalry.

Roosevelt yielded to Wood's superior combat experience to lead the regiment as colonel while accepting the post of lieutenant colonel. Thus began Wood's first experience with recruiting and training men for military combat. A recruitment center was set up in San Antonio, Texas, drawing on recruits from Arizona, Texas, New Mexico, Oklahoma and the Indian Territory. Roosevelt fought governmental red tape to secure the necessary, but very outdated, military equipment. Training proved a great challenge, as recruits were from a variety of backgrounds and experiences. Some enlisted from northeastern gentlemen's clubs, looking for adventure. Others were idealistic and patriotic volunteers. Few were seasoned men with military experience. After several weeks of developing some degree of organization, the regiment received orders to join the 5th Corps at Tampa, Florida. Traveling by train, the regiment, which included men, horses and equipment, arrived after three days in Tampa. Their arrival was greeted by military chaos, disorganization, and unpreparedness for the growing encampment of 25,000 men. Wood and Roosevelt did their best to take advantage of the situation. Their volunteer group organized their camp and set a daily routine of military exercises and study. The group was labeled the "Rough Riders." Over a week after their arrival, word was received for the regiment to prepare for departure from the docks located nine miles away. Eager to not be left behind the great adventure, Wood and Roosevelt performed some "rather lawless" acts to secure their regiment and the 5th Corps aboard confiscated ships. Roosevelt was left behind to deal with the aftermath of their actions.

Continued confusion among the military leadership caused delays, and even the arrival in Santiago Harbor produced chaos at embarkation. Wood and Roosevelt moved their troops in positions of their choosing, often at odds with their superior officers. Advancement toward military objectives demanded physical strength, disciplined troops and skilled leadership. The Rough Riders distinguished themselves in all areas.

In the infamous charge up San Juan Hill, Roosevelt's unit advanced without specific orders from Wood. After other examples of high command mismanagement, Wood initiated actions that helped capture the city of Santiago.

Singled out for his bravery and leadership, Wood was appointed brigadier general of the volunteers and named governor of Santiago de Cuba. After effective, but dictatorial control of the city, Wood initiated polices that improved the deplorable health, infrastructure, education and economy. Much of this was done in defiance of his immediate supervisor's orders and policies. After considerable subterfuge and acts of insubordination, Wood's supervisor was relieved of duty and Wood was named governor of Santiago Province. National publications in the United States glorified the heroics of Roosevelt's Rough Rider experiences and the effective local leadership of Wood in improving living conditions in Cuba. Changes in war department leadership and continued harassment by Roosevelt and Wood of their superior officers led to the appointment of Wood as military governor of Cuba. A national battle was raging in the United States over policy toward Cuba. Some sought annexation by the United States, while others sought Cuban independence. Those seeking independence won, but questions of timing and preparation were debated. Wood worked hard at implementing the movement toward independence and American business investment in the island. Political pressures prompted an earlier deadline for independence than Wood felt best. He relinquished his control of governorship on May 19, 1902.

Wood's appointment as brigadier general required congressional approval. Although he had much support, he also encountered attacks by political and military enemies. After contentious debate, Wood's appointment was approved in February 1901. There was much criticism nationally that a "personal physician, with no formal military training and only brief line experience, Wood's elevation over 509 other officers smacked of rank favoritism." It should be noted that the political manipulation of advancing in military ranks was a very common practice. Some received more public scrutiny than others. It should also be noted that these practices were in no way reflective of any more or less patriotic duty. These people were excellent military strategists and indeed patriots whose service to their country was beyond question.

Wood had established a very credible military and political influence. He also had created a reputation within the war department as an insubordinate self-promoter. In the fall of 1902, Wood and several other high ranking U.S. military officials were invited to attend the German Army maneuvers as personal guests of Emperor William II. Wood was impressed by what he saw, and would later use the observations as a guide for implementing his own military maneuvers. Meeting later with other European military and political leaders, Wood was warned about the German military intentions and cautioned to make the United States a militarily-prepared nation. With his travel report completed, Wood began to contemplate his next assignment. With his friend, Theodore Roosevelt, now president, lucrative business offers were more easily declined with the prospect of a successful military career.

Wood chose the Philippine Moro Province challenging command. With his experiences in Cuba, and the prospects for eventual command of the entire Philippine Division, it seemed like a good opportunity.

Two years of military campaigns, allowed very slow progress in province control; there were many casualties—military and civilian. Wood was also critical of his division commander. Wood was later appointed as commander of the Philippines Division. In March 1904, Wood was confirmed by the United States Senate as major general, in spite of strong political and military opposition. But many supporters wrote letters of approval, including General John J. Pershing, who was Wood's district commander in the Philippines.

The location of the Philippines was strategic for American interests in the Far East. Wood had to work at establishing military and political influence. The Chinese Boxer Rebellion had opened U.S. trade with China and allowed increased immigration to the United States. The Russo-Japanese War grew with threats of Japanese expansion into other Far Eastern territories. This created concern for a possible Japanese invasion of the Philippines. President Roosevelt asked Wood to prepare possible military action, however an eventual peace agreement was signed in 1908.

That same year Wood would end his term as commander of the Philippine Division. He would return to the United States with great hopes as the ranking army officer and in line for appointment as chief of staff. He made a tour of Germany on his return trip to review the German General Staff structure, in hope of solidifying desires to change the inefficient U.S. structure.

President Roosevelt was succeeded by William Howard Taft, who kept his promise to Roosevelt and appointed Wood as chief of staff in December 1909. As in the past, there was strong congressional opposition as well as support. Even after approval was won, Wood faced the physical challenge of surgery and a brief recovery for a brain tumor. The surgery, deemed successful, was actually incomplete and would later result in his fatality.

Wood's position as chief of staff afforded him great opportunities to make improvements in the nation's military organization and policies. It also gave him opportunities to gain new enemies, including a new U.S. president. The antiquated practices and need for reform "allowed an unconventional officer to become the prime mover for change. Yet those same conditions led Wood into explosive partisan politics, and ironically stymied and then destroyed a brilliant military career."

Wood became a strong advocate of military professionalism. Based largely on his personal observations and experiences, Wood and the younger military professionals were influenced by Emory Upton's, *The Military Policy of the United States*. Although written in the 1870s, it was not published until 1904. Upton argued that from past military experiences, the "Nation had paid the unnecessarily high price of wasted treasures and lost lives for its victories," by relying "too heavily in untrained volunteer soldiers." Upton called for "a small but highly trained, professional regular army to serve in peacetime. This group would be supported during war by "a reserve force already trained by regular officers." The prevailing antiquated system would overpower the need for change. Wood would prove to be the catalyst in moving the nation's military in a more progressive organization and philosophy. He would continue the "progressive efficiency reforms" begun earlier under War Secretary Elihu Root. As a fellow progressive with Roosevelt, Wood saw the need for U.S. military involvement in a changing world and the need for national military preparedness. Wood's effort to begin the changes started with the organization of the chief of staff position and command structure. These changes had the support of newly-appointed Secretary of War Henry Stimson.

Wood and Stimson faced major opposition from the entrenched military officers satisfied with their positions and philosophy. The new policies were challenged in Congress by supporters of both causes. Attempts were made to involve President Taft in correcting challenges to Wood's new policies by charging opposing high ranking officers with insubordination and "obstructionist tactics." After repeated and vicious personal battles fought in the political arena, Wood eliminated his main opponents by employing forced retirement and dismissal. As the chief of staff, Wood was now really the chief. Other areas of reorganization were needed in the War Department. Wood organized the general staff into four distinct and clearly titled divisions: The Mobile Affairs Division, The Coast Artillery Division, the Militia Affairs Division and the War College Division. Assistant chiefs of staff for each of the four divisions would handle "purely routine" matters. As a direct result, paperwork and the time it took to generate it was greatly reduced. Stimson estimated that the "government saved $245,000 a year in paperwork and overhead expenses."

The challenge of military mobilization was a serious problem. Before Wood's reforms, only about half of the called for forces were actually mobilized. The mobilized forces also reported to small and often unnecessary posts throughout the United States where only about an average of 700 men could be garrisoned. Wood and Stimson sought to eliminate about twenty-five posts over a period of time. Such a recommendation faced serious political opposition, especially from the affected states. The amendment was narrowly approved in the appropriations bill. But national political pressure against the bill and against Wood personally, caused President Taft to veto the amendment to the appropriations bill.

Another unpopular notion of Wood's was the universal conscription of young men into military service. A shortened length of service would hopefully gain some support. Congress failed to support any of the suggested reforms. The presidential election of November 1912, created an even more negative political environment for Wood. Democrat Woodrow Wilson had conducted a bitter campaign against Wood's close friend, Theodore Roosevelt. Many of Wilson's advisors had recommended ousting Wood as chief of staff. Several influential congressmen persuaded Wilson to keep Wood in the post, but his effectiveness was greatly limited. Lindley Garrison replaced Stimson as secretary of war, however a close camaraderie was established as Garrison favored many of Wood's suggested reforms.

In early 1913, two summer camps were established recruiting college students for four to five weeks of military training. From eighty institutions, 220 were trained in a highly successful experiment. Expanding the idea further, Wood, with approval from Garrison, established a training camp for Regular recruits in early 1914 at Plattsburg Barracks in New York. Gaining confidence, though uncertain about any congressional approval for monetary support, Wood began a series of public speeches to garner public support. In April 1914, Wood retired from the post of chief of staff. Faced with a decision for another military assignment, Wood declined the Southern Department involving conflict with Mexico and opted for the Department of the East command.

In the summer of 1914, Wood became aware of Germany's declaration of war against Russia. Wood now knew of the increased urgency to persuade the nation of needed preparedness activities. In this effort Wood gained strong congressional support. In December 1914, a National Security League was created to increase popular support. This was all in direct conflict with national policy and President Wilson's direction. However, Wilson was very supportive of the summer military training camps. Wood continued to make public speeches cautiously countering Wilson's reluctance for national military preparedness and an isolationist policy. In a speech in New York City, Wood crossed the line attacking Wilson. He quoted Revolutionary general Henry Lee: "A government is the murderer of its people which sends them into the field uninformed and untaught." That in itself was rather inflammatory, but Wood added his own commentary: "these words are absolutely true and these fake humanitarians who recommend that we shall turn the youth of this country into the battlefield unprepared are the conscious slayers of their people to an extent far greater than ordinary demands of war would render necessary." President Wilson demanded that Wood be reprimanded by Secretary Garrison. Since Garrison was much in sympathy with Wood's statements, he received only a mild rebuke. The war of words continued to escalate into 1915. Wood's support of the newly-formed American Legion caused Wilson to rebuke the general's aggressive actions. Wilson seemed to be in control of his desire to keep the United States from military involvement in the European war.

On May 15, 1915, the United States and its leaders would have a radical change in attitude towards preparedness with the sinking of the *Lusitania*. The summer camps that year saw great increases in recruits and an increasing emotional battle among those on either side of the preparedness issue. Wilson eventually became convinced of the political and military necessity of national preparedness. Increased naval and army expansions were reluctantly recommended. In May 1916, a National Defense Act was approved by Congress that established a National Guard reserve program. Still aware that insufficient numbers of volunteers could meet the demanded force numbers, great debates grew around the idea of universal conscription. Finally, when President Wilson asked Congress for a declaration of war on Germany in April 1917, he included a request for "universal liability to service." One month later, Congress passed the Selective Service Act, which called for a conscription of 500,000 men immediately and another 500,000 as needed, all remaining in service during the war effort. The emotional battles between the advocates and opponents of preparedness now reached an apex with Wood continuing to speak further inflaming these emotional battles.

Wood had unrealistic expectations that he would be named to command the U.S. military forces in Europe. Even as the highest ranking officer in the country, his enmity with President Wilson not only denied him leadership of the American forces, but he was denied any participation in the European military action. Wood was limited to training two divisions in the National Army. Wood's history of insubordination and unrestricted political involvement while in uniform caused his own downfall. He was transferred from command of the East Coast Department to the Southern Department.

When Wood discovered that General Pershing had been named to head the American forces in Europe, Wood held out hope that his past relationship with Pershing might motivate an offer of some major European command, but even that was denied to him for political reasons. As another act of humiliation, Wood was transferred to command the 89th Division at Camp Funston on the Fort Riley Military Reservation in Kansas. It would offer the least opportunity for continued political activity. Wood's training of these raw recruits was hampered by the customary problem of lack of equipment, clothing and weapons. However, improvising with crude substitutes, Wood trained the troops. It was also decided that Wood, as their commander, would not be going with them. The shock was great and another personal humiliation for Wood. He was assigned to train the 10th Division and remain at the Kansas camp.

During the training period, all commanders were ordered to visit the European theater to better assess the training needs. Wood saw this as a great opportunity to meet with Pershing and secure a combat command post. Making every attempt to dodge any meeting with Wood, Pershing denied any opportunity for Wood's participation in the European action. Pershing would explain his action with the following observation:

> With reference to a certain general, he is very hostile to the administration and has criticized the War Department freely.... His attitude is really one of disloyalty, in fact he is simply a political general and insubordination is a pronounced trait in his character. He is not in any sense true, and seemingly cannot control his overwhelming ambition for notoriety....

Upon his return to the United States, Wood stopped in Washington, D.C. to plead his case for combat command. He met with war and state department leaders as well as President Wilson. All listened politely, but rejected his request for combat leadership. At an ordered appearance before the Senate Military Affairs Committee, Wood again was very critical of the War Department general staff, the president and Pershing's leadership.

Such criticism certainly did not endear Wood to the national leaders. President Wilson expressed his feelings about Wood:

> Wherever General Wood goes, there is controversy and conflict of judgment.... I have had a great deal of experience with General Wood. He is a man of unusual ability but apparently absolutely unable to submit his judgment to those who are superior to him in command. I am sorry that his great ability cannot be made use of in France, but, at the same time, I am glad to say that it is being made very much use of in the training of soldiers on this side of the water.

Wood returned to the Kansas base and completed the five-month training of the 10th Division and saw it embark for European action without him. Frustrated and isolated by the national leaders, Wood needed to decide the direction for his remaining career.

Even though he had created a strong force of military and political enemies, Wood retained a large following of Republican leaders and the public who were supportive of him and offended at the snubbing he had received by the current administration. With encouragement of friends, and an appeal to his ego, Wood decided to officially enter the world of national politics. Three events shaped the path of such a political future: the Armistice in November 1918; the election of 1918 giving the Republicans control of the House and Senate; and the death of Theodore Roosevelt in January 1919. The Armistice provided the means for criticism of Wilson's vision for future American involvement in the affairs of the entire world. The 1918 election gave Wood a stronger congressional support for a presidential campaign effort. The death of Theodore Roosevelt gave Wood unshared leadership of the Republican party.

Wood formally announced his candidacy on January 7, 1920. He began an early behind-the-scenes campaign leading in the waves of popular issues concerning the "'Red Scare' and a movement to isolationism in the 'American 100-percenters.'" The issues were strong for Wood, but they were short-lived for the long campaign. Wood often campaigned in uniform, another sign of his rebellion and insubordination. The key party primaries were bitterly fought, especially in favorite son states where unusually high campaign spending was called into question. Wood came to the national party convention in Chicago with the largest number of delegates (not a majority), but many of those were *uninstructed*. Wood was also the party leader in many of the national polls. As Wood's strong campaign issues became less prominent, the party leaders sought a man who could unite the party and adhere to the advise of those leaders. Wood lost favor on both counts. After numerous inconclusive ballots, the party and delegates saw Wood's support dropping and looked elsewhere. After the tenth ballot, the nomination went to a surprised Warren G. Harding of Ohio. After the convention Wood gave strong support for Harding, especially after hints of Wood possibly being appointed Secretary of State. However, after the election, with Harding's landslide victory, Wood's position was much less in favor. After consultation with party leaders, Harding appointed Wood governor-general of the Philippines. A stunned Wood worked through his resentment of another apparent snubbing. He took the opportunity to provide patriotic duty to his country.

Actually, Harding had several reasons for the appointment of Wood to the post. First, in the Philippines, Wood was unable to maintain his high, national political profile. Second, with the past Democratic administration pushing for Philippine independence, Harding knew that Wood was in agreement with a more cautionary road to independence.

Wood began years of controversial control of the Philippines with strong opposition for those leaders anxious for independence. His policies even brought charges before Congress of expansion powers and denying current opposing political practices to continue.

Wood was strongly supported by the new president, Calvin Coolidge, in January 1924. Coolidge was in agreement with Wood's direction of a more cautious approach to independence of the Philippines. Wood worked tirelessly for another three years to provide for economic investment in the Philippines and secure a strong American presence in the Far East. He also worked to recreate the dictatorial leadership of the governor-general position he had in past Philippine leadership.

In May 1927, Wood returned to the United States for a conference with President Coolidge. While in the United States, the persistent struggles with the brain tumor forced Wood to consult medical help. It was found that the tumor had grown and was causing the physical difficulties that Wood had been experiencing for the past several years. The required immediate surgery proved to be more difficult than anticipated. During the seven hour surgery Wood lost consciousness, hemorrhaged and died in the early hours of August 7, 1927. Thousands gathered in Washington, D. C. to observe his burial with full military honors in the Arlington Cemetery, in a site reserved for the former members of the Rough Riders.

Conclusion

Leonard Wood made significant innovations in military training and preparedness that continue to the present day. His gospel of military preparedness led to the creation of the Reserved Officers Training Corps (ROTC) in many high schools and colleges. He also initiated the procedure for military basic infantry training camps for civilians. This training included use of the latest technology in warfare. His vision of a strong, well-trained standing army with Reserves and an efficient National Guard, are the standard practices for our nation's military today.

In spite of the strong support to name the newly-formed military base after Missouri native, General John Pershing, Wood's significant national contributions outweighed local pride in a native son. Because of his major emphasis on preparedness and military training, the base in central Missouri was officially named on January 8, 1941, after General Leonard Wood.

Chapter Three
1940s, WORLD WAR II

Introduction

By the early spring of 1941, construction continued at a rapid pace for the newly-named military base at Fort Leonard Wood, Missouri. Even before initial construction was completed, the arrival of troops, officers, office staff and grounds workers commenced. Some of the construction workers were now able to be housed in the new barracks, and were served cheaply priced meals at the new mess halls on base. Three shifts of workers provided the continuous, twenty-four hour efforts to meet construction deadlines that were primarily delayed because of weather. The logistics of providing adequate training would require carefully planned locations and facilities in a comprehensive base development plan. Safe firing ranges, physical training areas, and outlets for social activities were just as important as the barracks, mess halls and offices. Additional land would be required for training expansion. All of this activity was taking place many months before the United States was directly involved in an official capacity in the international war effort.

Ready for Training

With the national call for military preparedness, involving new base constructions and a national draft, events created the necessity of moving people to new locations for training. Initially these moves were done with inefficiency and confusion. Even at Fort Leonard Wood, things began with some degree of chaos. Recruits and officers began arriving at the yet to be completed base with orders that were often confusing and changeable. An example was found in the correspondence between Senator Truman and his friend, Lt. Col. Leo B. Crabbs. In late February 1941, Crabbs wrote the following:

Colonel Robenson at Corps switched my orders at the last minute and sent me here [Fort Leonard Wood] as he expected to be sent here as Post Commander. When he didn't show up I phoned Corps and asked who was Post Commander. They replied Colonel Robenson, so I phoned him and he said he was not coming, that I was in command…

I was the first officer here and the only Corps Area Service Command officer for the first two weeks, and was the Post Commander for the first month. Am now Assistant Executive and Morale, Recreation and Welfare Officer. As I have been here longer than anyone else and have had to settle a hell of a lot of problems I am a constant source of information for everybody.

Crabbs would also write about the confusion of early troop arrivals:

…eleven privates reported to me the next morning, having been sent with no notification preceding them. They had no rations or funds and no warm place to sleep. However they got fixed up until company "B" Sixth Engineers arrived shortly thereafter. I stuck my neck out for fifty-five meals for them.

In addition to the many buildings continuing to be constructed, training areas were also being established. Tank training sites, airfields, an engineer pontoon school, physical training sites, firing ranges and artillery ranges were needed. For a while, trains had to coordinate arrival and departure times with the close-by artillery range. Base security was established, but was very limited the first year. There was probably more concern with the actions of disgruntled and displaced local residents than with any foreign enemy. Office workers also began to arrive to assist with reception, processing and maintaining the mountains of paperwork required.

Infantry Divisions

The Sixth Division was to arrive on May 15, 1941. It arrived by train from Minnesota. It was to be joined later by the 72nd Field

Artillery Battalion on June 3. The battalion arrived by motor convoy from Fort Knox, Kentucky. During the war, a total of five divisions, the 6th, 8th, the 75th, the 97th and the 70th would train at the post. Early post commanders often served for only a few months before they were transferred. The initial mission of Fort Leonard Wood was the training of engineer replacement troops, Army ground forces and Army service forces units. The buildings and training facilities were designed to accommodate a maximum of 45,000 troops.

At the end of May 1941, the mass movement of troops began to settle in on the base. Scheduled for arrival was the 6th Infantry Division, 16,000 men; 72nd Field Artillery Brigade, 5,000; Engineer Replacement Center, 11,000; hospital force, 400; and Corps Service Command, 2,500.

An enlisted man's day was outlined as follows:

5:45 a.m.	Reveille; police barracks and grounds
6:15 a.m.	Breakfast
6:45 a.m.	Line up for drill with 10 minutes rest every hour
12:00 Noon	Chow; in barracks
1:30 p.m.	Drill
5:00 p.m.	Dressed in uniform for retreat
5:30 p.m.	Chow; men on their own
9:00 p.m.	Lights out; no talking in buildings
11:00 p.m.	Bed check

Men were off duty from Saturday noon until Monday Reveille. Wednesday afternoons were holidays except for recruits and their instructors. The eight-week basic training course qualified them for the Army. Most of the men were then selected for an additional eight weeks of specialized training.

Training was limited by the lack of adequate armament materials. The United States was ill prepared for the training of the large numbers of recruits. The nation had yet to change its industry and technology into a war production effort. A sense of urgency was not yet felt on a national level.

Declaration of War

The spark igniting a national commitment to war participation came as a result of the Japanese attack on the Hawaiian naval base at

Pearl Harbor on December 7, 1941. As President Franklin Roosevelt led Congress in a declaration of war the following day, the nation changed its priorities immediately. The draft soon created a large number of recruits. The attack on Pearl Harbor also motivated a mass movement of volunteer enlistees for military service. Industrial leaders changed their production lines for war armaments: tanks, planes, weapons, clothing, tents, etc. As men left their jobs to enlist, women took up the slack with dedicated patriotic service in industrial productions. All military establishments would feel the impact of the actual declaration of war with increased recruits and sense of urgency for training. The training would also take on a new sense of personal military combat preparation.

Base and Community Activities

The declaration of war had an immediate impact on Fort Leonard Wood. The base's security became a much higher priority. New entry gates were created with stricter requirements for entrance to the base. The base perimeter was also made more secure with the assistance of local citizens with specific, assigned duties. Paul Long, Waynesville resident, worked for the state highway department. His family's front lawn faced Route 66. They would sit on their front porch and watch the long processions of military vehicles. He was enlisted to provide base security immediately following the bombing of Pearl Harbor. Armed with his own personal shotgun, his assigned duty was to guard a Route 66 bridge near the base during the night with no means of communication. Later on, his family found the situation almost humorous—the idea of the Japanese attacking the rocky hills of Missouri and the picture of their father's "Barney Fife" character in national defense.

Traffic, already overtaxed on the limited, area roadways, experienced an even greater logjam. The traffic on Route 66, the only main road, increased dramatically. Military convoys arriving and deploying, and travel by construction workers, materials and equipment shipments, and daily staff workers saw the 7,000 vehicles daily average climb markedly. Although the new airfield and railway relieved some of the traffic, the highway congestion continued to increase. Traffic on base was also increasing due to machinery, military vehicles, and base worker vehicles.

With the area population increasing from just over 500 to over

50,000 within months, the community challenges and changes were overwhelming. Law and order deteriorated with only one town marshal and a sheriff in Waynesville. The base military police provided much needed assistance placing MPs in nearby towns. The local post office was inundated with incoming and outgoing mail service. Basic supply stores even as far away as Kansas City, Springfield and St. Louis were taxed to keep up with equipment demands.

Recreation

With the growing number of base personnel and the important but limited amount of leisure time, opportunities for recreational activities on base were created. From 1941-46, these recreational activities constantly expanded with the arrival of more troops. A field house for educational and recreational purposes was constructed. Sites for twelve base chapels were inspected for construction. Athletics comprised much of the recreational time, with a variety of sports and intramural competitions. An active base newspaper, the *Fort Wood News*, gave weekly reports of the base's recreational opportunities and results. Some of the early sports activities were boxing, baseball, basketball, billiards, tennis, bowling and track. In 1942, the first Golden Gloves teams on base were sent to tournaments in Kansas City and St. Louis. In 1943, the St. Louis Cardinals came to base and challenged the Base All-Star baseball team. The Cardinals won the game 3-0.

There were other recreational activities reported such as a camera club, literary societies, chess club, game nights, band competitions, plays, movies and dances. The demand for recreational activities on the war fronts also led to the creation of Special Entertainment Units. Two such units were trained at Fort Leonard Wood—Units Six and Seven. Their activities included a boogie-woogie piano player (with a heavy, portable 500 pound piano), short-wave radio, library, athletic equipment, game kit, hit movies and a limited post exchange. The first company of WAACs (Women's Army Auxiliary Corps) arrived on base March 20, 1943. The 153 women came from Des Moines, Iowa. They added to the interest of base dances and the ever popular pin-up contests.

The national USO (United Service Organization Inc.) organized stars and headliners to visit the military bases. Working with the Department of Defense since 1941, it has operated centers worldwide

1940s, World War II

depending upon contributions from individual and corporate donors. Primarily through its Camp Shows on military bases like Fort Leonard Wood, between 1941-45 the USO did 293,738 performances in 208,178 separate visits, entertaining more than 161 million servicemen and women.

Fort Leonard Wood was host to many of these special, well-attended events. Some of these stars are not well remembered today, but were very popular stars in the 1940s. Among these were the world-famous magician Cardini, actors Chick Chandler, Shella Ryan, Helen Cooper, Eddie Bracken, NBC news commentator W.W. Chaplin, Boxer Joe Louis and The Morgan Sisters Singers. In 1942, the St. Louis Symphony came to base for a performance with over 3,800 in attendance.

Off the base, the USOs created locations for soldiers on leave to have a place to go for entertainment and relaxation. Locally, there were USOs established in 1942 in Rolla, Lebanon and Waynesville. Rolla reported over 307,683 visitors the first year; 201,642 were soldiers. Usually dances were held with local women volunteering as dance partners. Special occasion parties, events and celebrations were also held at these centers. The existence of the rail lines on base also offered the opportunity for excursion trips to the major cities for sports and entertainment activities.

Cultural Diversity

Representative of the national population, the draft and the patriotic response answering the call to war, military bases felt the impact of cultural diversity in close quarters. For some areas unfamiliar with different cultures and races, there were unique challenges faced by the military. These challenges were present on the base as well as off the base among the local population. Such was the case with Fort Leonard Wood.

The *Fort Wood News* reported on August 6, 1943 that "110 Become Citizens in Mass Ceremony." These soldiers were able to secure citizenship through their military commitment. Pulaski County Circuit Court Judge, William E. Barton, administered the oath of citizenship. They were immigrants from a variety of countries: Austria, Britain, Canada, Czechoslovakia, Germany, Hungary, Italy, Mexico, Poland, Rumania, the Ukraine and Yugoslavia. Throughout the war effort, Fort Leonard Wood would experience other, smaller naturalization services. Many of the American soldiers had never encountered such cultural diversity. Some soldiers brought their own personal prejudices and suspicions from home to the base. There would be adjustments for the local community when encountering these internationals and for the internationals adjusting to Missourians.

At this time the challenge was even greater for blacks in the military. Even under the leadership of "Blackjack" Pershing, there was the impression that the African-American units in World War I were not successful. In the U.S. military of nearly half a million men, "there were only forty-seven hundred Negroes (the popular term of the time), two Negro officers and three Negro chaplains."

In the draft prior to World War II, African-Americans began enlisting in large numbers—nearly half a million by December 1942. Although the draft forbade discrimination "based on race or color in the selection of volunteers and draftees", the practice was very apparent in military organization and activities. It should be noted that the federal laws and Supreme Court rulings supported segregation at this time period. It should also be noted that even the inclusion of black baseball player, Jackie Robinson, did not occur until 1946, after the war. African American soldiers were "organized, trained, and housed in separate facilities apart from white soldiers." It was noted that "he [the African-American recruit] was quickly thrown into this furious activity to learn the skills of a soldier. In almost no time, as few as eight to twelve weeks, he was turned into an engineer soldier and shipped to an engineering unit. If he was white, he went to a combat unit, if black, he much more often went to a service or labor unit... Even in the few instances where black officers were senior to whites in the same unit, it was made perfectly clear in practice that no black officer, regardless of grade, would be superior to the most junior officer." Although this may have been the feeling of some African-American recruits, it is difficult to find actual documentation that supports such claims.

From 1940-45 the number of blacks in the U.S. Military increased from 5,000 to 920,000; the number of black officers from five to over 7,000. There were about 4,000 blacks stationed at Fort Leonard Wood in training and service units. Their housing was far from the parade grounds, requiring more time and effort to receive close-order drill and calisthenics. The classrooms were inadequate and their administrative buildings were inconveniently dispersed. As with other military bases, nationally, some local protests were made about the number of African Americans housed near small, white communities. In spite of the obvious segregated practices and community concerns, there were also instances of trying to improve community relationships. On November 20, 1942, a "group of fifty colored volunteers from the Quartermaster Detachment and the 49th QM Regiment became farmers for a day to help a widowed civilian harvest 2,500 bushels of corn."

Because of community concerns, special arrangements were made for recreational opportunities for blacks. After June 1942 the first designated black officers club was established in Building 2101, former offices for the Adjunct for the black engineer training group. A special program on base was planned for the "colored troops" on November 21, 1942 with novelist-playwright Langston Hughes making an appearance. A pre-Christmas dance for "colored" happened on December 19, 1942. Three hundred girls from Springfield, Jefferson City and Columbia were brought to base for the occasion. In addition, USOs for black soldiers were established in February 1943 in Rolla and Lebanon.

The continued discriminatory practices had a negative impact for the black soldiers and officers. In 1941 and 1942, the famous Tuskegee Air Flight program had produced nearly a thousand black combat pilots, "but by the spring of 1943 not single one had been sent into combat. One incident was reported in Louisiana when a group of nine black GIs had their train delayed for twelve hours. The only place they found to serve them food was in the kitchen of the station lunchroom. They were not allowed to eat in the whites-only dining room. A few hours later they noticed a group of about two-dozen German prisoners of war escorted into the dining room by two guards. Meals were served to them at tables in the dining room. The POWs ate, smoked, talked, laughed and had a good time. One of the black group members observed, "They were the enemies of our country, people sworn to destroy all so-called democratic governments of the world… What are we fighting for?"

When one black officer was asked at the end of World War II if he was going to reenlist, his quick reply was "Are you kidding? At least in civilian life the discrimination is not organized." Through the influential efforts of Eleanor Roosevelt and black civil rights leaders, the War Department on July 8, 1944 issued an order to all commanding generals that:

> all buses, trucks or other transportation owned and operated by the government or by a government instrumentality will be available to all military personnel regardless of race. Restricting personnel to certain sections of such transportation because of race will not be permitted either on or off a post camp, or station, regardless of local civilian custom.

It would not be until President Truman's Executive Order 9981 on July 26, 1948, that segregated practices in the United States Military finally ended.

Prisoners of War

Another group on base that created concerns were the large number of prisoners of war. Between 1942-46, nearly half a million prisoners of war (including Germans, Italians and Japanese) were held in some 500 camps in the United States. As many as 3,000 prisoners of war (POWs) were held at Fort Leonard Wood and satellite camps. In 1982 one of those German POWs, Fritz Ensslin, wrote of his experiences at Fort Leonard Wood. He served as a tank gunner in General Field Marshal Rommel's Afrika Corps. After capture by U.S. troops in May 1943, Ensslin boarded an American ship in Morocco. After thirty days at sea he landed in Norfolk, Virginia. He then boarded a well-secured rail car. After a two-day trip he arrived at Fort Leonard Wood at midnight, June 1943. He and the other POWs were met and escorted by a large force of well-armed security guards and taken to Camp 1.

The camp was well-lit, ringed in barbwire and had many manned observation towers. Ensslin was directed to the barracks that each housed fifty men. He was surprised by having his own bunk, having spent years sleeping in a tank or on the ground. At 1:00 a.m. he received what he described as a "dream meal." He joked with his fellow POWS "If we had only known, we would have sneaked across earlier instead of fighting until we ran out of ammunition." After the meal, he was allowed to sleep until noon.

Ensslin received his GI uniform, which was dyed blue with large

POW letters printed on the back, knees and rear-end. He worked with the others to set up general housekeeping and grounds improvements. With the others, he did his own laundry and again was surprised by the use of a special barracks with sinks, toilets and bathing facilities with hot and cold running water any time of the day. No smoking was allowed for the first six weeks.

After the first weeks of adjustments, volunteer work on base began. Thirty men would be needed for laundry, but only ten of the 1,000 POWS volunteered. After the first day, the laundry volunteers reported that there were "at least 100 girls there in all shades and colors." Many others wanted to volunteer. From then on, the laundry was operated twenty-four hours a day with thirty POWs on each shift. There was one guard for every ten POWs during work. The guards were often abusive, while Ensslin described the American women and engineers as treating the POWs as co-workers.

Other regular work performed by the POWs on the base included mess hall, shoe repair, road construction, grounds maintenance and tree removal. Other work details included pin setters in the bowling alley, firing of furnaces in officers' quarters, garbage removal, incinerator workers, office cleaners, painting, carpentry and cabinet makers. The POWs earned ten cents an hour for eight hours a day, paid in five and ten-cent money certificates. POW camp stores were established for purchase of personal items and luxuries. When not at work, the POWs were allowed to write, read, paint or carve. Sports and group artistic activities were permitted. Tutoring by the better-educated and skilled POWs was also permitted. They also had access to American newspapers and magazines to keep up with war news and other international events. There was even a short-wave radio in their camp, where on occasion they could hear one of Hitler's speeches.

On one occasion over abuse by a guard, the POWs went on strike. During this time, the POWs received bread and water for meals while the POW camp commander worked to iron out the problem. The base had become so dependent on the POWs that a strike was felt across the entire area.

In the spring of 1944, Ensslin, in a group of thirty POWS, was allowed to leave base and work on a Missouri farm. In spite of the growing hatred for anything German in the country, these POW groups were treated well. They helped with harvesting cotton, sugar beets, peanuts, corn, spinach, melons, potatoes, etc. His last job on base was working in the mess hall serving meals to soldiers returning from the front, awaiting discharge from the Army. In the spring of 1946, the POWs began to be returned to Europe through France. There they were to spend their time in forced labor. He evaluated his experiences as an American POW by noting the adherence to the Geneva Convention in the handling and treatment of POWs.

Soldiers and Local Relations

As with many American military bases, the relationship with the local communities was a mixture of problems and benefits. Activities of some soldiers led to an increase of local criminal incidents. Many of these resulted due to the abuse of alcohol and the lure of illegal and immoral enterprises. Residents and veterans recalled trailers and motels just outside the base gates, where prostitutes operated. Gambling sites were available to take advantage of the recently-paid soldiers.

There were many more opportunities for positive experiences within the area around Fort Leonard Wood. Many soldiers were able to take advantage of the available outdoor exploration. Local churches benefited from the number of visiting soldiers and offered hospitality and prayers. Base choirs performed in the churches. The community was invited to attend theater productions and watch special intramural sports events on the base. In June 1943, major flooding occurred in central Missouri. Over 1,000 post troops joined in battling flooding at nearby Bagnell Dam.

Training Accomplishments

The primary purpose of the military base was the training of troops for combat service. As the war progressed, additional troops were needed and replacement troops were in demand. Adjustments in training were needed due to experiences on the field of battle. Then Army Chief of Staff General George Marshall created procedures for the standardization of training practicing throughout the country. New scientific technologies provided improved training and field operation. New technologies in weaponry, vehicles, building materials and radar created the opportunity for more advantageous military operations. Personal accounts recall the experiences of recruits at Fort Leonard Wood in receiving the personal and unit basic training.

1940s, World War II

Roger O. Austin provided a written record of his training experiences at Fort Leonard Wood two years after the attack on Pearl Harbor. He was a young man raised on a farm with plans to attend Cornell University. His early efforts in ROTC accelerated his advancement for enlisted training. His departure by train from the East Coast was adventurous and educational. At each stop he recalled the great care given to all of the railway equipment—maintained now only by women. The arrival in the St. Louis Union Station was impressive. From St. Louis, they rode the train backward all the way to Fort Leonard Wood, because there was no place to turn around upon arrival. On unloading, they were immediately taunted by graduating troops now ready to embark for deployment.

Taken by truck to the barracks, bunks were selected and their gear was stowed. After a quick cleanup, they headed to the mess hall, where to their delight seconds were encouraged. The next morning the 5:00 a.m. routine began. After the required policing duties were completed, the mess hall was ready for breakfast. Following that, medical screening was begun. A detailed physical included dental checkups and screening for venereal disease. After that there was a series of marching, calisthenics and jogging. Along the routes on base, there were new discoveries, such as German prisoners of war being used as groundskeepers. Austin related that the July and August heat in 1942 Missouri was especially miserable for all and cold showers were welcomed.

In addition to the routine physical training activities, there were five, ten and even twenty-five mile marches often accompanied by an ambulance. The recreation hall was used for special classes and training films. Some of the courses were found interesting and challenging. Obstacle courses were demanding, but an essential part of their training. As preparation for engineering assignment, wooden bridges were built over the streams and rivers. The hardest duty described was "digging those damned fox holes." The rocky Missouri soil was even more of a challenge because of the many big machines that he said had crossed that same land daily.

Austin noted that rare weekend passes provided a welcome break from the confines of Fort Leonard Wood. A trip to Jefferson City and a visit to the state capitol were very enjoyable. A good meal and a comfortable hotel room bed were also welcome respites.

Upon nearing completion of his basic and engineering training,

Austin had several continuing training options. He chose the Army Specialized Training Program. He summed up his Fort Leonard Wood experiences as follows, "My training at FLW was tough, thorough and interesting. It was an experience I could never forget and I would miss the men in my barracks. What I would never miss was the heat, the dust and the Great Piney River."

Accomplishments
National

No history of the great accomplishments of Americans in World War II can be complete without recognition of the contributions made by the civilians to support the war effort. Just as the military was mobilized to train and equip the vast number of troops, civilians also mobilized in the support of the war effort. The supply of cotton and wool were needed to produce "more than 64 million flannel shirts, 165 million coats and 229 million pairs of trousers." Accordingly, the War Productions Board mandated changes for civilians. For the men, "victory suits… with cuffless trousers and narrower lapels" were ordered; for the women "reductions in the amount of cloth allowed for shorter, pleatless skirts, rising several inches above the knee and the creation of a new two-piece bathing suit." Home victory gardens and metal collection drives also assisted with the support of the nation's war efforts.

Among the greatest accomplishment of civilian support was the change in manufacturing productions from peacetime supplies to war material needs. It was only after the attack on Pearl Harbor that American manufacturing leaders took seriously the urgency of the need for sudden changes to be made. Detroit was transformed from the world's foremost manufacturer of automobiles to the leader in production of war materials. General Motors began making planes, anti-aircraft guns, aircraft engines and diesel engines for submarines. Ford now was producing bombers, jeeps, armored cars, troop carriers and gliders. Chrysler was building tanks, tank engines, army trucks and mine exploders. "The industry that had once built four million cars a year was now building three-fourths of the nation's aircraft engines, one-half of all tanks, and one-third of all machine guns." The *New York Times* boasted, "At the end of the first year of her intensified war effort, the United States was turning out more war materials than any other country in the world."

Even the Russian leader Joseph Stalin, November 30, 1943, praised the manufacturing efforts of the United States:

> I want to tell you, from the Russian point of view, what the President and the United States have done to win the war. The most important things in this war are machines. The United States has proven that it can turn out 8,000 to 10,000 airplanes per month. Russia can turn out, at most, 3,000 a month. England turns out 3,000 to 3,500… The United States, therefore, is the country of machines. Without the use of these machines, through Lend-Lease, we would lose the war.

It was also noted that "Between 1940 and 1945, the United States contributed nearly three hundred thousand warplanes to the Allied cause. American factories produced more than two million trucks, 107,351 tanks, 87,620 warships, 5,475 cargo ships, over twenty million rifles, machine guns and pistols; and forty-four billion rounds of ammunition."

A major factor in the ability to produce such accomplishments in the midst of war operations was due in large part to the role of women in manufacturing. Over nineteen million women were employed, constituting one-third of the entire labor force. Many of the residents surrounding Fort Leonard Wood were among those who served at the home front supporting the national war effort.

Local

In less than five years, over 300,000 individuals were trained for military service at Fort Leonard Wood. Averaging 60,000 soldiers being trained per year, the story of the base rising from wooded, mountainous and rocky land to the sprawling acres of training sites is an amazing credit to the commitment of a nation and patriotic individuals. Civilians became soldiers and combat-ready in as few as eight weeks of basic training. New facility construction was ongoing. Local citizens adapted to the many changes in their area and the level of activity that thousands of workers and soldiers had created. Federal expenditures in the area had been significant and beneficial for many local residents. Businesses had been created and expanded to meet growing needs. Transportation in and out of the immediate area had been improved with part of Route 66 now expanded to four lanes. Train and air transportation improvements to the base helped to ease some of the traffic problems. As eager as the entire nation was for the conclusion of the war effort, no one was prepared for the effects of its conclusion.

Post World War II and Base Inactivation

With victory over the Axis powers by the end of 1945, the United States had to deal with many major national and international issues. These issues had immediate impact on Fort Leonard Wood and the surrounding areas. The national economic adjustments from wartime to peace would prove challenging on personal and corporate levels. Along with the budgetary issues was the concern for price controls, especially of rationed items. Employment for the many men and women serving in the armed forces was also a concern. In June 1944, a new G.I. Bill of Rights passed the Senate and House of Representatives with no opposition. It would ease the unemployment situation and provide educational and occupational training to create and improve skills for employment. The bill would see over two million veterans receiving college and graduate school educations at a cost of $14 billion. There was an urgent need for new civilian housing with the returning forces. Families would be reunited, and the role of women in the workforce would change dramatically. As these challenges were met on a national level, the areas around Fort Leonard Wood would also undergo significant changes.

Although the title of "fort" implies more permanency than "camp," the need for financial expenditures required the closure of many such bases in the nation. Even with a Missourian now in the White House, the state would directly feel the impact of such base closures. On January 21, 1946, a confidential memorandum was sent from Army headquarters to the deputy chief of staff for service command at Fort Leonard Wood. It announced the plans for the base closure around the end of March 1946. The Army ground forces units would be moved or inactivated. Prisoners of war would be moved to other locations. An exact date was yet to be determined.

In February, on the base there was a strong emphasis for re-enlistment of service men and women. The base newspaper, *The Fort Wood News*, announced the first article about the base closure on March

8, 1946. All training would cease as of March 23. The engineering center would close on March 31.

There were now many technical issues to be settled on the base. The 70,945 acres on the base were valued at one to five dollars an acre, but were deemed as unsuitable for agricultural, industrial or residential use. The worth of the land was determined to be only for use as a valuable watershed. There was discussion about the railroad spur to Newburg, pipelines, recreational areas, utilities and facilities. Plans were also discussed to allow limited areas of continued training during the summers by the state national guard and army reserves. A small caretaker unit would remain to maintain some of the buildings and facilities.

The impact outside of the base was more detrimental, as thousands of civilians lost their post jobs. New, local businesses (good and bad) would cease to exist. The impact of the lost income spent locally by the military and the generated tax base was significant. Local and state appeals for a change of orders for the closing of the base came too late and would have little effect. Local schools and other civic organizations would also undergo major changes in numbers and operations. The effects would be felt most in Waynesville, Rolla, Lebanon and other nearby local communities. As the citizenry watched the departures of troops, officers, prisoners of war, and office personnel, there were mixed reactions. The war had ended and their husbands, wives, sons and daughters would be returning home. Just as the rapid influx of over 30,000 workers flooded the area in the spring of 1941 with so great an impact, the equally rapid departure of the 60,000 military personnel had an even greater local effect. The local citizenry had performed miraculous accomplishments to allow for the effective training of victorious soldiers. But there was also the sense of hopelessness as to the future of the communities now abandoned by the closure of the base at the conclusion of the international conflict.

From the summer of 1946 through 1950, there was comparatively little activity on the now inactivated army base. However, one new, major event was led by recent residents Bob and Dormalee Morgan. They came to Pulaski County in 1948 from Oklahoma. Bob was a WWII veteran who worked cattle. His contract for cattle grazing in Oklahoma was terminated in 1948. Bob worked to arrange with the Jarboe Commission Company of Tulsa, Oklahoma, to lease Fort Leonard Wood from May 1948 until August 1950. The Morgans lived in a house bordering the base and ran the operations with help from family and friends in the cattle business. Many fences were built and pastureland was determined. Many of the eventual 3,000 head of steers were brought to the base by truck from Texarkana. Roundups were conducted with many ranch hands ready to lead cattle drives on some of the main gravel highways to pens on the base. The herds were then transported by rail to packing houses and other pastures out of state.

Some of the local farmers were able to sell their steers for inclusion in the roundups. One of these was the Laughlin family. Bill recalled that many of the Oklahoma cowhands had trouble working with the Missouri land and terrain. He was hired with other locals to work the herds. His family also brought 6,000 steers to Fort Leonard Wood in 1946. They continued to have cattle on the base until as late as 1963.

During this same time in the summer, the base lands were shared for two-week, scheduled, annual training for the national guard.

Conclusion

By the end of the 1940s, Fort Leonard Wood was mainly a site for cattle drives instead of military training. Lassos replaced rifles and cowhands replaced recruits. The base landscape and facilities were now covered with weeds and saplings and in deteriorating condition. The area around the base was just as depressed with the loss of employment, income and population. However the area could be very proud of the major contributions it provided to the victory in WWII. The nation had joined together in a unified effort to provide the training, equipment and technology in order to defeat a powerful, international enemy. Personal and family sacrifices were made by U.S. citizens for the aid of others across the oceans. These sacrifices were made by a nation that had no plans for gaining the territories of the Allied or Axis nations.

What may have appeared to be the ending of significant national activity would be changed in rapid fashion just months after the next decade began because of another international military conflict. Fort Leonard Wood would not only revive, but find its permanent place in operations and importance.

Photographs

Staff of the 1st US Volunteer Regiment (the Rough Riders) in Tampa, Florida, ca. 1898. Pictured r-l are Theodore Roosevelt, Leonard Wood, and Civil War Confederate Genberral Joseph Wheeler, with Taylor MacDonald to the far left and Maj. Alexander Oswald Brodie next to him. Photo by Underwood & Underwood. Courtesy of the Library of Congress, LC-DIG-ppms-ca-37597.

General Leonard Wood, Army Chief of Staff, on horseback, riding in front of a reviewing stand at the Riggs National Bank during the first inauguration of President Woodrow Wilson in Washington, D.C., 1913. Photo by Bain News Service. Courtesy of the Library of Congress, LC-DIG-ggbain-11338.

The History of Fort Leonard Wood, Missouri

General John J. Pershing, ca. 1920. Born in Missouri, the base was nearly named in his honor before the decision was made in favor of General Leonard Wood. Photo by Theodor Horydczak (1890-1971). Courtesy of the Library of Congress, LC-H823-1534.

Group at the gravesite of Leonard Wood, 1929. Photo by Harris & Ewing. Courtesy of the Library of Congress, LC-DIG-hec-35502.

Photographs

Bad weather challenges early base construction efforts, ca. 1940-41. Courtesy of the Engineer History Office.

Early base construction, ca. 1940-41. Courtesy of the Engineer History Office.

The History of Fort Leonard Wood, Missouri

Early base construction, ca. 1940-41. Courtesy of the Engineer History Office.

Terrain also challenged early base construction efforts, ca. 1940-41. Courtesy of the Engineer History Office.

Photographs

Early base construction, ca. 1940-41. Courtesy of the Engineer History Office.

Early base construction, ca. 1940-41. Courtesy of the Engineer History Office.

Installing utilities and communications in the early base construction, Courtesy of the Engineer History Office.

Early base construction, ca. 1940-41. Courtesy of the Engineer History Office.

Photographs

Early base construction, ca. 1940-41. Courtesy of the Engineer History Office.

Early base construction, ca. 1940-41. Courtesy of the Engineer History Office.

Early base construction, ca. 1940-41. Courtesy of the Engineer History Office.

Skilled carpenters and carpenter's assistants help in the early base construction, ca. 1940-41. Courtesy of the Engineer History Office.

Photographs

Early base construction, ca. 1940-41. Courtesy of the Engineer History Office.

Early base construction, ca. 1940-41. Courtesy of the Engineer History Office.

The History of Fort Leonard Wood, Missouri

Base water tower, ca. 1940-41. Courtesy of the Engineer History Office.

Much blasting was needed for clearing roadways in early base construction, ca. 1940-41. Courtesy of the Engineer History Office.

Photographs

Building complexes begin to take shape in the early base construction, ca. 1940-41. Courtesy of the Engineer History Office.

New roadways were needed on base and in the surrounding area, ca. 1940-41. Courtesy of the Engineer History Office.

Early base construction, ca. 1940-41. Courtesy of the Engineer History Office.

Photographs

Railway construction from Newburg was a major undertaking through the wilderness and mountains, early base construction, ca. 1940-41. Courtesy of the Engineer History Office.

Constructing the railway to the base, ca. 1940-41. Courtesy of the Engineer History Office.

Early base construction, ca. 1940-41. Courtesy of the Engineer History Office.

Early base construction, ca. 1940-41. Courtesy of the Engineer History Office.

Photographs

Early base construction, ca. 1940-41. Courtesy of the Engineer History Office.

Early base construction, ca. 1940-41. Courtesy of the Engineer History Office.

Base water tower. Courtesy of the Engineer History Office.

Numerous bridges were needed for the roadways and railway, early base construction, ca. 1940-41. Courtesy of the Engineer History Office.

Photographs

Early base construction, ca. 1940-41. Courtesy of the Engineer History Office.

Good weather and pre-fab building construction made up for delays, early base construction, ca. 1940-41. Courtesy of the Engineer History Office.

The History of Fort Leonard Wood, Missouri

Early base construction, ca. 1940-41. Courtesy of the Engineer History Office.

Early base construction, ca. 1940-41. Courtesy of the Engineer History Office.

Photographs

Early base construction, ca. 1940-41. Courtesy of the Engineer History Office.

Early base construction, ca. 1940-41. Courtesy of the Engineer History Office.

A multitude of workers cooperated in the construction of barracks, early base construction, ca. 1940-41. Courtesy of the Engineer History Office.

Photographs

Aerial view of the barracks area, 1940s. Courtesy of the Engineer History Office.

Aerial view of Fort Leonard Wood, 1940s. Courtesy of the Engineer History Office.

Photographs

Early base construction, ca. 1940-41. Courtesy of the Engineer History Office.

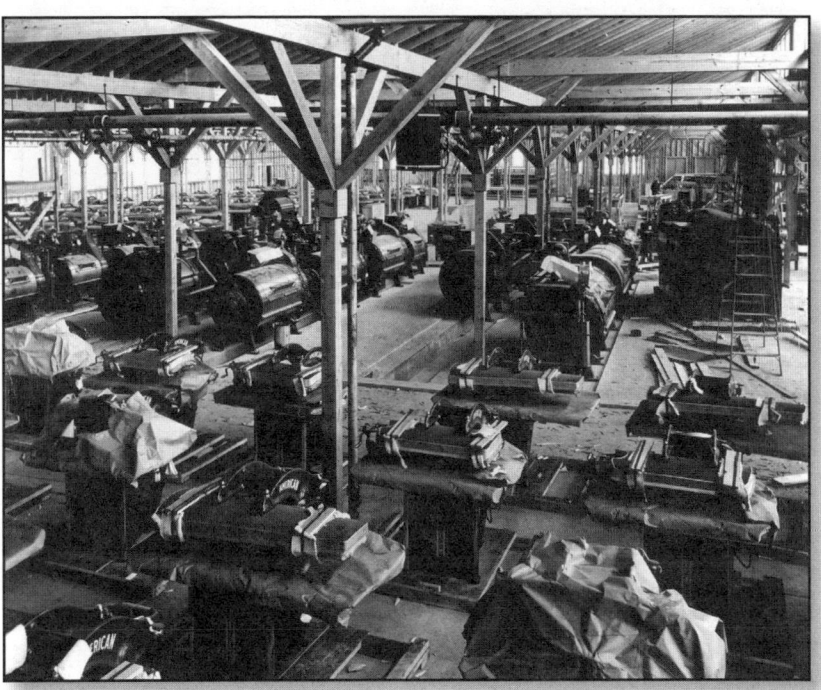

Early base construction, ca. 1940-41. Courtesy of the Engineer History Office.

Early base construction, ca. 1940-41. Courtesy of the Engineer History Office.

Aerial view of Fort Leonard Wood, 1940s. Courtesy of the Engineer History Office.

Base dance during the 1940s. Courtesy of the Engineer History Office.

Base dance during the 1940s. Courtesy of the Engineer History Office.

Photographs

Base dance during the 1940s. Courtesy of the Engineer History Office.

Base dance during the 1940s. Courtesy of the Engineer History Office.

Postal service and storage provided challenges during the early base construction, ca. 1940-41. Courtesy of the Engineer History Office.

Photographs

The most effective form of transportation on base during early construction was horseback, early base construction, ca. 1940-41. Courtesy of the Engineer History Office.

Major A.F. Ofederstrom, Asst. quality manager, Hospital Area No. 1. Courtesy of the Engineer History Office.

Construction worker families lived in cramped quarters, early base construction, ca. 1940-41. Courtesy of the Engineer History Office.

Peeling potatoes for the evening meal, 1940s. Courtesy of the Fort Leonard Wood Base History Office.

Photographs

Base cooks work in the mess hall. Courtesy of the Fort Leonard Wood Base History Office.

Chow time in the 1940s. Courtesy of the Fort Leonard Wood Base History Office.

Air service was very limited during early base construction, ca. 1940-41. Courtesy of the Engineer History Office.

Early base construction, ca. 1940-41. Courtesy of the Engineer History Office.

Chapter Four
1950s, KOREAN WAR

Introduction

At the beginning of the 1950s, America was poised to thrive on a peacetime economy and the development of a strong middle class. Many G.I.'s had taken advantage of the G.I. Bill, enabling them to receive a higher education. Housing developments began to spring up, offering affordable housing for the masses. Automobiles, household appliances and recreational opportunities were provided for access by the middle class. The United Nations had been created in the hope of avoiding further international military conflicts, however, an apparent path to peace and contentment in the United States was shattered on June 25, 1950 in a faraway locale—Korea.

Declaration of Involvement

After World War II, the United States became concerned with the expansion of the Soviet Union in Eastern Europe, with the insurgency in Greece and takeover of the Czech Republic. There was also concern for the People's Republic of China's desire for control in the Far East. These concerns were expressed in the phrase, "The Cold War," that would continue for over forty years. When 75,000 North Korean forces invaded the southern peninsula (in what was to become South Korea) on June 25, 1950, the United States was caught by surprise. The invasion by the communist-controlled forces was more than a simple civil war or border dispute—it was seen as a first step toward communist aggression in order to control the Far East. When the Soviet Union boycotted the United Nations Security Council emergency meeting, it allowed the United Nations to act as an international body to militarily defend the newly-formed South Korean government. Although the United States would provide 88% of the 341,000 international armed forces, it would again need to put greater emphasis on preparing troops for combat action.

Military involvement for the United States began in July 1950, as the first American troops were deployed to South Korea. The Korean involvement was called a "police action" in an effort to avoid the possibility of a third world war. The people of the United States would receive far less publicity and information about the Korean conflict than during the recent history of WWII. Perhaps this was purposeful to avoid the title of a third world war and defend the limited action military approach. Although the United States had had the previous experience of war in the Pacific theater, the efforts on the Korean peninsula brought new challenges. Training and technologies of the recent world war would need to be refined and enhanced for the new, international police action.

Reactivation

Official word was announced in August 1950, that Fort Leonard Wood would be reactivated as a military training base beginning in September 1950. That meant the base would need to be ready for the arrival of 20,000 troops within one month. Fortunately, some earlier work had already begun in preparing the base, as the initial American troops were getting ready to be trained to be deployed to Korea. But major preparations were needed to rescue the base from five years of any regular inactivity.

In July 1950, a young recruit, Jerry Obrey, was among the first arrivals on the base to prepare for its probable reactivation. Obrey remembered that the base was overgrown with brush, grass and saplings. There were even small trees in the doorways of the barracks. Obrey spent many days digging ditches and recalled that "he never felt so sore." He spent eighteen months at Fort Leonard Wood and recalled many memories of his experiences there. For example, just outside the base entry gates were trailer homes where long lines of soldiers waited to see strippers inside for the dime entry fee. His weekends included trips to St. Louis and Columbia, Missouri. He and his buddies dated girls from Columbia College. They were not allowed to pick up the girls in their cars, but arranged to meet them at a local lake. On the base he remembered his long nights in the winter on guard duty at the motor pool. His toes were often frozen and required slow warming in the nearby huts. Despite these hard times, he remembered them fondly.

As troops began returning to the area in large numbers, local

1950s, Korean War

business and community activities increased. Vacated post jobs on the base needed to be filled by local residents. Reba (Long) Parker worked as a secretary in the base headquarters in 1953. She remembered the facility as having no insulation for protection in the winter months. During the summer, floor fans were used and the workers needed to take salt tablets because of the heat.

One of the major challenges to the life on base was the deplorable housing available for soldier families. The existing housing was poorly renovated WWII construction. This was not unique to Fort Leonard Wood—it was a scene on most military training bases. The limited, available housing off the base was not much better. Congressional hearings were held to investigate the numerous complaints.

Recreational opportunities and various means of transportation were needed, however, assuming a continued, temporary activation status for the base, no new construction was planned or funded. All too aware of the local, negative impact of the inevitable future deactivation, community and legislative efforts began to work for the goal of a permanent status for the base.

Training Activities

Korea was a different kind of military conflict with new objectives. However, the tactics, techniques and equipment were the same as in training the troops in WWII. In the past, training was done for multiple divisions. Now, training for replacements was done primarily by one division—the Sixth Armored Division. Training was also accomplished at an increased rate, with the establishment of a reception center on the base. Now more recruits could be processed more quickly, getting them ready for their basic training assignments. New army personnel from thirteen Midwestern states would now be processed and trained. Fort Leonard Wood continued as either a vital basic training or engineer training center.

Experiences in Korea required new training techniques for the recruits. Weather and terrain necessitated specialized training and equipment. North Korean mass advancements and blowing trumpets were tactics different from WWII. A cooperative effort of mixed United Nations forces required improved organization and communication.

Training activities on the base continued at a regular pace for replacement troops. For the young recruits, Fort Leonard Wood held memorable experiences—good and bad. Such was the case for Dale L.

Geise. In 1950, Private Geise arrived at the base for training as a part of the 982nd Engineer Battalion. He recalled his first meal in the mess hall, remembering the coal-burning ranges and the oak food boxes filled every day with chunks of ice. He remembered falling out and marching to the firing range, leaning into a cold November rain. He, too, remembered the days of "cleaning a ditch followed closely by a road grader that filled it again. Army discipline." He had fond memories of Cpl. Roskelly at daily, morning inspections asking recruits to give him fifty push-ups because "maybe he had seen a little brown shoe polish in the eyelets of boots, or perhaps a quarter tossed on the stretched khaki blanket of a bunk didn't bounce up and flip over." He admitted that it all caused him to become a better soldier. He cherished the "memories of friends made at a time when we became soldiers together, and were part of something bigger than we could imagine."

James Jaeger was a recent high school graduate who went immediately into basic training at Fort Leonard Wood in the summer of 1956. He recalled the strenuous activities of the eight-week basic training program. On base Jaeger attended movies and drank beer at the PX. One trip off base was to tour the Rolla School of Mines. His memories of Fort Leonard Wood were "not real good," but he felt that the experience did provide very good training for later active duty. After basic training, he went to the Aberdeen Proving Grounds for ordinance training.

Community Changes

The area immediately adjoining the military base was rapidly developing and outside the jurisdiction of Waynesville. The area was known as St. Robert, named after the local Catholic church that was named after a Jesuit Cardinal, St. Robert Bellamine. In October 1951, the area was officially chartered as a village. With the future base permanency, the area would develop into a small city.

A number of local residents indicated that St. Robert was actually established over the liquor-by-the-drink issue. The more family oriented, church-going community of Waynesville refused to approve the issue.

A longtime resident of the area, Joe Miller, arrived in 1950. He came by Pullman car arriving in Newburg after midnight. He then took a bus to Fort Leonard Wood. He had already had a tour of duty in WWII. He was single and called up for what was assumed to be deployment to Korea. But after sixteen weeks on base, he was informed that he would not be going to Korea. He was assigned special schooling and worked in supply and training for two years. After his service, he chose to stay in the area and worked in marketing and insurance. He married a local girl on February 14, 1953. They have been married for 63 years (2016). Joe worked with the eventual Fort Belvoir transition with activity in Washington, D.C. He indicated that Fort Leonard Wood made the area more of an international community.

Cold War Training

Although on July 27, 1953, a truce was finally reached in the stalemated military conflict, the truce was very shaky with suspicions held on both sides. A continued United Nations military presence would need to be maintained on the Korean peninsula. Continually trained American combat forces would be needed to replace those rotating out of military duty. Other areas of concern necessitated a significant, trained and armed military force to be ready for rapid deployment. The hopes of a small, peacetime, military force were abandoned for a large standing army ready to answer the Soviet and Chinese threats.

Soviet political influences and military activities began to spread to eastern Europe, Africa and even in the western hemisphere, particularly Cuba, Central and South America. Continued tensions in divided Berlin caused concerns for additional, military preparedness. The development of Soviet nuclear weapons and the launch of the Soviet satellite, Sputnik, in 1957 created widespread panic in the United States about new advances in weaponry and space technology with its potential military implications. A space and arms race began with heavy scientific and economic demands for both nations. These events would also affect Fort Leonard Wood's military training and technology required now for offensive and defensive preparedness.

Base Permanency

With the drawing down of troops being trained for the Korean Conflict, word began to spread about the likelihood of another deactivation of Fort Leonard Wood. There were some immediate down-sizing effects on the base following the conflict, as would be repeated other times in the future following major military conflicts. The hardships of the last deactivation on the community were to be fought in a very organized and effective manner. Local and statewide community leaders created the Committee of 50 led by Dru Pippin of Waynesville, to "Make Fort Wood Permanent." They created an effective and widely-circulated brochure entitled, "Why Fort Leonard Wood Should Be Made Permanent." It indicated the strong evidence of advantages for continuing its operation: unlimited land availability for the base expansion, location isolation and security, abundant water supply, variable terrain and seasons for training, and accessibility to three large cities. The national interstate highway system was underway, offering much needed connectivity in the Midwest, particularly with Route 66. Working closely with the committee was the Missouri congressional leadership. Especially helpful was U.S. Congressman from Missouri, A.S.J. Carnahan. In early 1956, Carnahan (father of later Missouri Governor Mel Carnahan) guided the progress of HR Bill 9893 through the Armed Services Committee and into congressional and military debate. The committee was very effective in lobbying congressmen and Defense Department personnel.

In January 1958, the Honorable Dewey Short, assistant secretary of the Army, announced at Fort Leonard Wood before a gathered assembly of Army personnel and civilians, that the base would remain and be made permanent. A twenty-year plan was introduced to replace the "hastily constructed temporary buildings by modern permanent buildings." A total cost estimate was put at two hundred million dollars. The announcement had an immediate impact on the base, the community and the state. The base would see ongoing, new construction and expanded training facilities. The community would have the confidence of economic security and stability. The state would see the continued importance in its place in national defense.

The past congressional hearings on military housing conditions prompted the priority of expenditures toward housing. There was also a tension created between the base and the community. The more housing provided on the base meant fewer housing needs off the base. That same tension continued for over four more decades.

On the base, construction projects began immediately. Ground was broken for much needed base housing. On February 27, 1958, an initial thirty-three family housing units were begun in the Capehart Housing Project. In August 1958, ground was broken for major permanent troop housing that replaced many of the wooden structures hastily built before WWII. In 1958, construction was also begun on the

Regimental Training Areas, labeled the "rolling pins." Among the other projects planned were a new base hospital and headquarters facilities.

The decision for permanency of the base also had an impact on the local communities. In the past, with the temporary status of the base, there was reluctance to assist with the local burdens of an increased population. But now the federal government offered substantial assistance for much needed community projects. In Waynesville, the government supplemented the building of a water and sewer system. With the influx of military families, the area schools were overcrowded and now in ill repair. Temporary school buildings and a new, million-dollar high school were constructed.

Fort Leonard Wood permanency also had an impact on the entire state. Enterprises from around the state flocked to the base area in numerous business interests. Fort Leonard Wood became the sixth largest city in the state.

Conclusion

With the ever-growing presence of the Cold War, the United States became more concerned about military preparedness. The fears of two world wars and a prolonged "conflict" in Korea, still unsettled, led the United States to learn the value of ongoing military training and the ability for rapid mobilization to potential war fronts. Fort Leonard Wood continued to provide military training with the latest technology available continuously for the next sixty years.

Chapter Five
1960s, VIETNAM WAR

Introduction

With Fort Leonard Wood gaining the status of a permanent military base, its significance as a military training site increased. The major federal expenditures on the base and in the community were also indicative of the base's importance for national defense. Although on a national level no official declaration of war existed, there were vital areas of international conflicts that had the potential to easily and rapidly escalate into military action.

Berlin, Cuba and Other Areas of Crises

In the aftermath of WWII, many countries with colonial dominance in areas on other continents were facing economic and political pressures to divest themselves of continued control. This provided the opportunity for the Soviet Union to begin aggressive ventures to fill the void in those previously dominated areas of the world. The immediate impact was experienced in the neighboring countries of Eastern Europe. Soviet occupation of those countries, in opposition to United Nations concerns, created an atmosphere for potential conflict.

The divided, occupied city of Berlin continued to be a source of great, potential military concern. The Truman Doctrine prevented the Ageaen peninsula from falling into Soviet control. In Latin America and Cuba, in the late 1950s and 1960s, Soviet influence was felt through economic assistance as well as some military assistance to revolutionaries seeking independence from foreign colonialists for their countries. Such was the case in many countries in South America. The same Soviet influence was also present in the emerging independent countries in Africa. In 1965, 20,000 U.S. troops invaded the Dominican Republic to counter the communist-led uprising to overthrow the country's elected leadership. In 1967, the U.S. military provided limited assistance in the Congo (Zaire) to provide logistical support during a revolt.

The Cold War between the western world and the Union of Soviet Socialist Republic (USSR) was being realized on a global scale. The communists conducted a successful takeover of mainland China in the late 1940s. The French abandonment of the small country of Vietnam in 1954 allowed for growing communist influence in the Far East. It was at this time that the Eisenhower administration developed a concept of what was called "containment," leading to the later idea of the "Domino Theory." If the Soviets controlled one country in the region, others would soon fall into their domain. Accordingly, Eisenhower authorized military aid and the deployment of military advisors to Vietnam in 1956 to train the Southern Vietnamese troops to assist them in countering the aggression of the communist, North Vietnamese forces.

A major, Cold War flashpoint occurred in Cuba during the fall of 1962. Soviet missiles had been discovered on the island with the potential to strike major cities in the United States. In a dramatic showdown, the Kennedy administration established a quarantine of the island and required the dismantling of the Soviet missiles. Successful in their avoidance of a nuclear confrontation with the Soviets, the Kennedy and Johnson administrations would begin to assess the possibility of stopping further Soviet expansion in the Far East by becoming more involved militarily in Vietnam.

Escalation in Vietnam

Communist insurgents increased their activities in Southern Vietnam through weapon supplies and bombing attacks. The first U.S. casualties resulted from these attacks in 1959. The new Kennedy administration pledged loyalty and support to the new South Vietnamese President Diem. The U.S. Air Force began a bombing campaign on North Vietnamese Army supply route areas with Agent Orange—a defoliant. This led to a widespread and controversial practice of chemical warfare. Questions arose over the leadership of President Diem by U.S. politicians and Buddhists. There were charges of his misuse of two billion dollars of U.S. aid, as well as a practice of religious intolerance in South Vietnam. In 1963, Diem was overthrown with U.S. approval and replaced by General Nguyen Khanh. This same year, President Kennedy was assassinated and Lyndon Johnson became president.

On August 2, 1964, three North Vietnamese ships allegedly fired torpedoes at a U.S. destroyer in the international waters of the Gulf of Tonkin. Five days later, Congress, in the Gulf of Tonkin Resolution, authorized President Johnson to "take all necessary measures to repel any armed attack against forces of the United States and to prevent further aggression." This allowed the U.S. to wage all out war against North Vietnam without an actual congressional declaration of war.

The first actual combat troops were sent to Vietnam to protect military targets, such as airfields and bases, that were under Viet Cong attacks. After Johnson's election as president in the fall of 1964, Operation "Rolling Thunder" began in 1965. Beginning in February, sustained massive bombing took place over North Vietnam for the next three years.

The first significant American combat troops arrived in South Vietnam in 1965. An escalation of American combat troops deployed to South Vietnam reached 200,000 that year. An additional 206,000 troops were requested by American military leadership in February 1968. The demand for trained troops and replacements created increased activity at Fort Leonard Wood, but also generated increased opposition from some national leaders.

Basic Training

In June 1960, a young, single Missourian, Clark Hale, entered his nine-week basic training at Fort Leonard Wood. He had been raised on a farm and was accustomed to wearing boots and coping with the summer heat of Missouri. He realized that some of the recruits from areas in the east coast had never worn boots and were not prepared for the heat. This resulted in making them even more uncomfortable throughout the training process.

Hale recalled loss of sleep and being treated as though he were "hardly human." He remembered the barracks as having coal-generated heating and hot water supply. He also remembered wearing his Sunday dress khakis in the mess hall line. The coal stoves inside the mess halls produced a steady stream of coal dust outside that dust settled on his dress khakis. Friends would simply blow it off and others would try to wipe it off, resulting in the dust being patted in making them harder to clean later. In addition to the standardized training activities, Hale experienced guard duty and fire patrol while on the base. He had no experiences off the base during his time there. Going later to Fort Sill for artillery training and eventually to Germany, he was grateful for the good training he had received at Fort Leonard Wood.

Walt Badshaw was a member of the Kansas National Guard in Manhattan, Kansas in 1967. He did eight weeks of basic training at Fort Leonard Wood beginning April 1, 1967. He recalled the packed lines on base for their shots and watching some of the men faint. They were lifted off to the side of the line as the shots continued. He had a lenient unit led a by a black drill sergeant who hated running and did not require that of his soldiers as much as other units did. He was allowed to attend a camp baseball game one evening if he donated blood. His unit was on standby for possible deployment to Detroit, Michigan, for race riot control. He mentioned that in Kansas he flipped burgers at the Burger Barn. He was designated as a cook and spent eight more weeks at Fort Leonard Wood in Advanced Individual Training in cook school. He was joined by a number of National Guard members from Minnesota. He was allowed a car kept off base by the main gate. He made trips to Waynesville for beers and one weekend trip to Columbia, Missouri, where his girlfriend's family had moved. His experiences in Vietnam were constant reminders of the good training he received at Fort Leonard Wood.

As experiences in Vietnam grew, the training for this type of jungle combat was aided by creative techniques at most military bases, including Fort Leonard Wood. In the late 1960s, a replicated Viet Cong campsite was constructed with simulated Viet Cong soldiers. Trainees would patrol the wooded area of the base, unaware of the campsite location. Once the base was discovered, the trainees were asked for observations providing tactical information. The campsite also served as practical preparation for pre-deployment as one of six training stations. The first were examples of booby traps and a punji pit (a large camouflaged pit filled with sharpened bamboo stakes). The second was various sniper nests. The third was the actual campsite. The fourth was a solid rock bunker. The fifth was a collection of mines, barrels and ammunition boxes. The sixth was a rice trap. Observations were reported by the trainees and evaluated. Field training also included the capture, search and processing of a Viet Cong prisoner.

Remembering the experiences of the Viet Cong campsite training was a young soldier from Topeka, Kansas, named Steve Williams. He was drafted February 6, 1969 and arrived at Fort Leonard Wood March 6, 1969. He was just a young kid and scared to death of his drill sergeant. He recalled being moved from his wooden barrack to a new brick barrack. His eight weeks of basic training provided life-saving lessons for his experiences in Vietnam. Of real value was his training in low crawling, survival training, working as a unit and developing use of all his senses. He served as a "pointman" in Vietnam.

Engineer training on the base was largely unchanged with the need for construction of roads, base camps and airfields. There were some improvements in equipment used.

As the demand for additional troops was approved, more men of draft age were called into military service. More families became directly involved with interest in the war's progress when actual films of battle were brought into American homes, via television, on a nightly basis. Because of the continued length of the war, more replacement troops were needed in order to maintain the increasing troop levels. These demands caused shortening of some basic training from eight- to six-week periods on many of the military bases, including Fort Leonard Wood. The height of training at the base occurred in 1967, when some 123,000 soldiers received training in basic combat and engineer skills. That was over double the original, intended, training plans for the base.

Edsel Matthews served as a special services officer at Fort Leonard Wood from May 1965 to June 1967. Excerpts from his recollections are as follows:

> One of the most interesting and flamboyant personalities during that time was my First Sergeant, Herman "Big Train" Jackson. He had spent 26 years in the service when I arrived. He was one of the original Afro-American soldiers to be integrated in the U.S. Army.
>
> His ability to get things done was uncanny and unique. Spending most of his service at Ft. Wood, he had the trust of the military personnel and civilian population alike.
>
> Upon my arrival, I was informed that in the last year Big Train had run off five first lieutenant special service officers who had come in and talked down to him and failed to use his assets.
>
> When I arrived in May, I had just finished being

part of a coaching staff that had won the Missouri State basketball championship in the largest [state] classification. Big Train officiated that final game. Consequently, we had a common bond from the beginning. I told him he knew more about the service than I ever wanted to know. I said, "Train, you tell me how things are to operate, I'll ask questions, and we will make this work." He could tell more stories and share more experiences than anyone at the fort. By following that leadership model, we were highly successful and got things done that put Ft. Wood on the map.

A professional standard golf course was created on the base when Big Train traded two dozen golf balls for 200 basic recruits for three days' work on improvements on the golf course. A big Chamber of Commerce Gold Day invited golfers from Lebanon, Springfield, Rolla and Waynesville to play the course with the base officers. All were impressed with the course that kept its high standards for two years. Big Train also helped Matthews put together a base basketball team that won the All Army Championship in 1966. Big Train continued officiating basketball games and held popular clinics to train many young basketball officials. Matthews concluded his thoughts by saying "Big Train was truly a legend and will always be remembered by those who knew him for his outstanding and community service."

James Wallace was a young recruit who came to Fort Leonard Wood in January 1967 for his basic training. He remembered many other recruits as "young, right out of high school and a fairly rowdy, high-spirited bunch." Nicknames were a big thing for them that fortunately only lasted through basic training. Wallace's nickname was "Wally Gator." He recalled one fellow from West Plains who had an impairment on the right side of his face that made it difficult for him to speak clearly and to eat. The West Plains recruit was nicknamed "El Lavio Lips" by an insensitive group of guys and had great difficulty with the physical fitness demands. He had another recruit take the physical test for him and wound up with the best grade of anyone in the company.

Wallace took a lot of abuse from much of the company. They would hide his prize camera and get him into trouble that at times would require him to be assigned to latrine duty. When time came to ship out, everyone was taking souvenir photos. El Lavio Lips got his prize camera and had every one line up for photos in various positions and at various camera settings for 15 to 20 minutes. After he finished, he laughed and said, "Got you guys. There's no film in the camera." El Lavio Lips was actually the first person in the company called to the CQ room to receive his orders and the first to leave. Wallace related that "as he walked down the stairs with his duffel bag, he shouted loudly enough for everyone in the barracks to hear, 'You sonsabitches can all line up from here to hell and kiss my ass. Bullshit!' That was the last time I ever saw him, but I have never forgotten him."

With the need for trained soldiers rapidly increasing, Fort Leonard Wood continued in its role of basic training and engineer training. The base was assuming more responsibility for advanced engineer training than its counterpart in Fort Belvoir, Virginia. The growing urbanization of Fort Belvoir's location near Washington, D.C. limited its activity in ordinance training. Fort Leonard Wood's vast acreage allowed for greater freedom in engineer training and the use of new technologies.

Military Draft

The military recruiting system was under much criticism throughout the mid and late 1960s. This was because many believed that minority groups and the less advantaged bore the majority of the burden of military service. Deferments of draft status were seemingly reserved for the rich and influential, although many young men received deferments for education, ministry, medical needs and family status. Compounding the shifting national attitude toward the Vietnam involvement was a series of dramatic national events. In 1967, veterans staged well-publicized, anti-war rallies and prominent leaders, such as Dr. Martin Luther King, Jr., spoke out publicly against the war. 1968 proved a pivotal year in national politics with presidential and congressional elections. Societal questions about the war and national leadership were evolving in national concern with the quagmire in Vietnam. Well-respected television news commentators, such as Walter Cronkite, announced their negative evaluation of the war efforts, especially after the TET offensive in the early days of 1968. In March 1968, President Johnson announced his decision not to run for reelection; in April, Dr. Martin Luther King, Jr., was assassinated;

in May, the Paris Peace Talks began with little success; in June, Senator Robert F. Kennedy was assassinated; and in August, the Democratic National Convention in Chicago was disrupted with all of these events televised to the world.

In November 1968, Richard M. Nixon was elected president. He campaigned on the platform of bringing the Vietnam War to an end. He would push for controversial military campaigns in the Southeast Asian peninsula and serve as the new target for anti-war demonstrators.

In 1969, the program of "Vietnamization" was begun with the goal to turn over combat responsibilities to the South Vietnamese. That spring, a national draft lottery was conducted to more fairly recruit new troops among those who had registered for the draft.

From 1965-73, 1.7 million were drafted and inducted. Of those, approximately 40 percent were deployed to Vietnam. An increase in "draft dodgers" was noted with many finding refuge in Canada. That same year, massive anti-war rallies were staged and televised in Washington, D.C. Other new recruits created increased interest in the Navy, Air Force and Coast Guard in an effort to avoid deployment to Vietnam by the Army and Marines.

Base Activities

All of the previously mentioned national events played out while the training continued at Fort Leonard Wood. As the base was expanding its training load in 1967, 25,000 reservists were facing call-ups for active duty. Tony Dow, the actor on the popular television show "Leave It to Beaver," began his basic training at Fort Leonard Wood. The USO continued its active role in entertaining the troops and showing support for the war efforts. The popular singing group, "Up with People" made several appearances on the base throughout the 1960s.

In the fall of 1966, Si Zentner and his orchestra played Big Band music for several performances on the base. Anita Bryant, popular entertainer, also appeared for performances on the base that year. At Fort Leonard Wood during 1967, the list of entertainers appearing on the base included Hans Conreid, Dorothy Lamour, Buck Owens, the St. Louis Symphony, Phyllis Kirk and Johnny Cash and June Carter.

The USO continued its support with appearances on the base that year by Bob Hope (April 19), Gina Lollobrigida and Phyllis Diller. Other base activities in the 1960s included several successful blood drives sponsored by the American Red Cross. Many sports competitions continued with ad hoc teams as well as company organized teams. In the 1960s, six-man football teams were organized for competition.

In January 1968, even with the base's record training numbers, then commanding officer General Walker ordered budgetary cutbacks at Fort Leonard Wood. Local newspapers reported those local residents killed in action in Vietnam. There were also reports of the base stockade escapees, AWOLS and the increased crime on the base and in the community. It is likely that the national events and changing opinions of the soldiers toward the Vietnam War affected attitudes regarding military training at Fort Leonard Wood.

National attitude towards the war and the military was experienced by the example of a young recruit from Oklahoma, James Hromas. After arrival at Fort Leonard Wood for basic training, he excelled in marksmanship on the rifle range. Receiving the highest score in his company, he was rewarded with a three-day pass. Wanting to surprise his family and his girlfriend, he left the base by bus and arrived home in Oklahoma. For the entire time of his visit at home, all were fearful that he was AWOL, finding his story of a three-day pass unbelievable.

Community Impact

As Fort Leonard Wood was reaching its numerical apex in training soldiers during the Vietnam War, the fort also had a significant impact on the local communities. There was a direct economic impact on nearly 50,000 people and indirectly on about the same number. In 1967, people working or living on the base in military service totaled about 35,000. There were about 4,000 civilian employees of the Department of Defense and about 8,000 military dependents. Payments to military and civilian personnel at the fort were reported in excess of $65 million for the fiscal year ended June 30, 1965. Including other government purchases, total expenditures at the fort exceed $100 million annually.

When most central Missouri counties experienced a decrease in population, the counties along Route 66 from Fort Leonard Wood to St. Louis all experienced population increases. In the immediate area around the fort, there were 4,000 school-age children added to the public education systems. Over $1 million in federal aid came into the communities for building programs and employment of new education staff. With no overnight housing provided on the base, local motel business was booming.

Bill Morgan's family lived on the oldest farm in Missouri, dating back to 1828. The base bordered on their farm property. Bill was stationed at Fort Leonard Wood from January 14 through August 13, 1960. Bill and fellow resident Bill Debo recalled the square in Waynesville being filled on Saturday afternoons with busloads of soldiers on weekend leave from the base. Every fifteen minutes a Williams bus shuttle from the base would unload on the square. Nearby were parked Cadillacs, with prostitutes inside soliciting the soldiers. Also remembered, just off the town square, was the Mitchell Glass Company that made yellow tinted and aviator glasses and binoculars.

David Johnson lived on a farm bordering the west side of Fort Leonard Wood. He was stationed at the base in the summer of 1969. He recalled seeing AWOL soldiers walking on the road by his house. His family or the neighbors would call the base MPs or the local law enforcement to pick them up.

Accomplishments

In the fall of 1968, Fort Leonard Wood was declared the "nation's largest U.S. Army Training Center." By 1969, Fort Leonard Wood's population ranked as the seventh largest city in the state, behind St. Louis, Kansas City, Springfield, St. Joseph, Independence and Columbia. Throughout the 1960s, an average of 94,000 men were trained each year—49,000 in basic combat skills and 45,000 in engineer and combat support specialties. The highest levels of training occurred in 1966 and 1967, during which almost a quarter of a million men were trained at Fort Leonard Wood. With that increase, new facilities were built for housing and training.

The construction cost on the base during 1966 and 1967 was over $24,000,000. Much of the construction was for permanent housing. For a while, over 2,500 recruits were housed on the base in tents. Since the reactivation and decision for permanency in 1956, more than $86,000,000 was spent for construction of permanent facilities.

Conclusion

The great turbulence in the American society of the late 1960s was felt in the support of and opposition to the Vietnam War. As the war escalated, drafted troops pushed the base to its training limits and local communities felt the impact of the increased numbers. Fort Leonard Wood had become the nation's premier training center. The positive benefits on the local communities were offset by negative impacts that would find unfavorable light in the 1970s.

Chapter Six
1970s, Vietnam War

Introduction

As a nation grew weary of the Vietnam War and the national discord being experienced, events added fuel to the fire of discontent. President Nixon and Secretary of State Kissinger were working publicly and behind the scenes to bring it to a conclusion. Even with American troop levels in Vietnam drawn down to 280,000 in 1970, the riots and deaths at Kent State that year created further anti-war, anti-authority sentiment. In 1971, the secret Pentagon Papers were published, revealing the nefarious path of U.S. military involvement in Vietnam. Troop levels, in 1972, were reduced to 70,000. That same year the Watergate breakin occurred, threatening political turmoil.

With Nixon's overwhelming presidential victory that year, more serious peace talks began. An official ceasefire was signed in 1973, and the end of the draft was announced. In 1974, the war was briefly renewed and Nixon was forced to resign because of the Watergate investigation.

Under President Ford, the last American troops evacuated Saigon, leaving South Vietnam to be overtaken by the North Vietnamese. Ford was also facing the strain of a faltering economy with very high inflation rates. The flood of returning troops from an unpopular foreign war, and those returning "draft dodgers" now pardoned, created high unemployment and resentment among the troops who were feeling coolness from an apparently ungrateful nation.

Demobilization

The nation's changing role in Vietnam and the evolving culture of the U.S. had an impact on military training bases throughout the nation, including Fort Leonard Wood. With the beginning de-escalation of troops, the training pace began to slow down. Defense spending had created a growing challenge to the national economy. Even while actively engaged in the Vietnam War, in March 1970, the Pentagon announced cuts for 371 bases, involving 93,900 jobs, to save $914 million. The U.S. Army experienced the deepest cuts from its wartime peak of 1.6 million and 19 divisions in 1968 to 785,000 soldiers and 13 divisions in 1975. These levels would be maintained until 1990.

Fort Leonard Wood was scheduled to lose 254 military jobs, saving $1,800,000. It was reported that, "A high frequency radio facility will be phased out. Reorganization of the strategic army reserve and a decrease in the training workload are responsible for the changes that will occur by June 30."

Just several years after the peak training numbers of the late 1960s, reports began to filter out in the national press, which had a negative impact on the military base and the surrounding communities. While these reports had a local impact, it was emblematic of reports on a national scene. The *Kennett Missouri Weekly* reported on April 4, 1970, a summary of national impressions about Fort Leonard Wood, drawn largely on the late 1960s experiences. Quoting irate parents in the Springfield area, the summary stated that, "the base prescribes too much work and too little medical attention." The *Washington Post* reported that Fort Leonard Wood "is an army base that has an unprintable description among troops who have taken basic training there." *Esquire* magazine reported that the Missouri base ranks just slightly behind Fort Polk as the most miserable base in the nation.

In early 1970, a serious outbreak of meningitis infected 53 soldiers on the base. Between 1967 and 1969 there were 68 cases reported. By March 1970, the situation created concern statewide. An elementary school teacher was barred from her school and classroom after her Parent Teachers Association learned that she had visited her husband at Fort Leonard Wood.

A congressional delegation toured the base for one day and had mixed conclusions. Three Missouri congressmen on the House Armed Services Committee gave the army a clean bill of health; however, two Missouri congressmen, Senator Thomas Eagleton and Representative James Symington, were responding negatively to the deaths of some of the soldiers infected. Representative Thomas (Tip) O'Neill from Massachusetts viewed Fort Leonard Wood as a "modern-day

Andersonville," insisting that the national guardsmen from his state not be sent to Missouri until improvements were made. New vaccines, developed in February 1970, at the Walter Reed Army Institute, promised improved treatment and prevention of the meningitis threat later that year.

In the fall of 1975, a series of newspaper articles reported on serious crimes that took place on the base. Headlines reported the following: "DIs [Drill Instructors] charged with blackmail and bribery," "Court martial for 9 sergeants," and "Private charged with selling LSD." At the same time there was a newspaper report headlining that, "Five Schools from Ft. Belvoir To Relocate At Fort." In spite of the reporting of the small minority of problems on the base, there was enough confidence in its future to prepare for its expansion.

Base Activities

Even though the public relations aspect of the base was suffering due to negative reports, the Pentagon was reassuring in its support. In April 1970, it announced plans for $1,946,000 of new construction on the base. Among the new permanent facilities constructed in the 1970s, were a post exchange, commissary, credit union, the Truman Education Center, bachelor officers quarters, officers club and the Specker Barracks.

In August 1972, Fort Leonard Wood welcomed its first black general, Brigadier General Edward Greer, as the new deputy commanding general of the post. In 1973, the U.S. Army Training and Doctrine Command (TRADOC) was established on the base. In 1974, an interservice training review organization was established. In 1978, the first gender-integrated basic training was established at the base.

With growth in recruits and training leaders, the expanded housing on the base created a rapid growth in the base schools. Teachers for the new grade schools were recruited from the local areas. One of these teachers was David Suits, a veteran of the Vietnam War now settled with his wife, Gaytha, in Laquey, Missouri. He had just finished getting his master's degree from Central Missouri State University, taking most of his classes at the Fort Leonard Wood Truman Education Center. From 1975-78, he taught fifth and sixth grades at Williams Elementary School on the base. He has the following recollections:

All of my students had parents in the military, mostly NCOs [non-commissioned officers] stationed at the fort. Some of my students also knew each other from previous schools attended together in other countries. The families were very supportive of what was going on at school, attending parent conferences and PTA meetings.

Something interesting we found out was that at the local stores there were often others there that spoke languages other than English, including Spanish, Vietnamese and Korean.

Suits also taught a variety of psychology classes at the Truman Education Center for Drury University.

Community Relations

Just as the base itself had undergone negative reporting around the nation, the areas around the base were also seen in a negative light. The *Jefferson City Post Tribune* noted that "crime and violence are also getting out of hand." The *St. Louis Globe Democrat* editorialized on September 1, 1972 in an article entitled "Clean Out Viper's Nest."

When Fort Leonard Wood opened in January 1941, it was probably a close race to see who would arrive first—the drill instructor on the field or the prostitute outside the gate.

These vipers' nests of prostitution, gambling and illegal liquor have too long perverted the borders of the army base and surrounding countryside.

Current battle among the Pulaski County mobsters is only the latest skirmish in a three-decade-old war which has pitted the underground against the law, even hoodlum against hoodlum.

Because the base is a government installation, federal authorities have a deep obligation to commit manpower to the fight against the camp followers who prey on young soldiers. State and local authorities must also join the battle.

Much of the cream of American youth is sent to

the 71,000-acre fort in the Ozarks. It is intolerable that these young men, and the law-abiding citizens who live in the area, should be exposed to the flesh-peddling scum who infest the region.

Local, state and federal authorities should intensify efforts to erase this filthy smudge from Missouri.

Less than a week later, the same newspaper reported on a speech by Missouri Attorney General John C. Danforth prepared for delivery in Columbia, Missouri. Excerpts from the speech included:

The people of the [Pulaski] county have not desired good law enforcement. They have not wanted wide open vice to be curtailed. They have apparently taken the position that crime can be profitable to a community.

What is happening has all the earmarks of a gang war waged over the control of prostitution and gambling.

If a community is willing to tolerate crime, then it will have crime. It is just as simple as that.

The article also reported that "the army has not placed a nightspot off limits in the Ft. Wood area in at least three years and none are off limits now." An advisory board did report the finding to General Bradley and resulted in prompt directives placing several establishments off limits to military personnel.

It must be noted that this situation was not unique to Fort Leonard Wood. Nearly every military base in the United States suffered from some of the same challenges. Because of Fort Leonard Wood's unique location and status as the largest military training facility in the nation, it would be singled out for evaluation by soldiers, parents, and their political representatives.

A retired Colonel (not wanting his name printed), recalled some of the negative aspects of the community for the soldiers in the area. He came to Fort Leonard Wood in July 1967 as an experienced pilot from action in Vietnam, where he flew generals in the combat zones. He worked two years as publicity officer and at the flight school. He was witness to some of the illegal operations at the Top Hat Club, with slot machines and prostitutes. He returned to Fort Leonard Wood in 1980-84 as a full Colonel. He remembered fondly the work of Frank O'Malley, who was a "doer" as the Special Services director, always working for the troops. He enlisted big name entertainers and shows for the base. He created a new golf course and riding stable. In the 1980s, the Colonel saw a much more family-oriented community that treated recruits and their families better than any other post where he had been stationed. He retired in the area teaching physical education.

In spite of the later negative reporting, the 1970s began with promising community optimism. The *Fort Gateway Guide* reported in January 1970, "Business and professional men, laborers and the general citizenry see continued progress, with much building now underway, despite the 'tight money' situation which prevails all over the United States."

Keith Pritchard is a successful banker in Waynesville and also serves as a Civilian Aide to the Secretary of the Army, Missouri-West. His father moved their family from St. Joseph, Missouri to Fort Leonard Wood on December 31, 1952. He served the Army there until his retirement in 1982. His father tested new equipment and tools for the Army Tool Committee. He gained a national reputation for his work and evaluations. After retirement, he was a Waynesville city councilman and became mayor from 1988-92. Keith also recalled that he and his family helped his father to build houses on their farm. One house was rented to Bud Walton, then owner of Ben Franklin stores #1-6, which eventually became WalMart. His father turned down offers to invest in the new supermarket.

LeRoy Mellows was a black Sergeant Major who served at Fort Leonard Wood for several periods from 1972-86. He was part of the 2nd Engineer company and the 3rd Battalion basic training company. He was the first enlisted commandant, replacing a Lieutenant Colonel. He was the supervisor for the infamous "million dollar hole." He married a local German woman and liked the area very much. Much like the previously mentioned "Big Train," LeRoy was a well-respected referee of many area sporting events. He personally experienced no racial trouble, even with warnings from other blacks.

Military Interventions

After withdrawal from Vietnam, the U.S. military had limited activities in Lebanon and Zaire (Congo). Continued unrest in

Latin America caused our military to be prepared for invasion and involvement to counter the communist assistance of so-called independence movements. Of major concern in the late 1970s, under President Jimmy Carter, was the major unrest in Iran that led to the capture of the American embassy and the kidnapping of American hostages. American troops began to prepare for possible military involvement in and against Iran.

Conclusion

The horrors of a prolonged and unpopular war in Vietnam had come to an end. Confidence in America's leadership, political and military, was very low. The possibility of any further international military involvement was limited by low national morale and economic challenges at home. The returning vets faced an apathetic welcome from the American public. The nation was facing critical gas shortages with long lines at filling stations and the national economy was being crippled by high inflation rates.

Fort Leonard Wood, not exempt from problems, remained the premier Army training center in the United States with the possibility of expansion during a time of military financial cutbacks. Furthermore, with the threat of Iranian radicals holding Americans hostage, unrest in African countries, Latin American countries and the Middle East, the 1970s ended with the prospect for peace unlikely for the 1980s. The future for Fort Leonard Wood, however, appeared to be bright with potential growth and stability for the community.

Chapter Seven
1980s, Cold War

Introduction

The United States military was undergoing strategic changes in the 1980s. The 1970s had seen the conclusion of the Vietnam War with ensuing cutbacks in manpower and budgets. This new decade found the country with concerns that involved more limited military actions. On April 26, 1980, President Carter used six U.S. transport planes and eight helicopters in an unsuccessful rescue attempt of the Americans held hostage in Iran. President Reagan used the military in advisory roles to quell an attempted coup in El Salvador. Military forces, throughout the 1980s, were also sent to other places including Lebanon, Honduras, Chad, Bolivia, Honduras, Panama, Columbia and Peru. President Reagan also called on the military for an invasion of the island of Grenada. In 1989, President George Bush used the military as peacekeepers in the Philippines. That same year, President Bush sent U.S. military forces to Panama to protect American lives and interests and to bring the rebellious General Noriega to justice. These actions required a constant state of readiness in order for the U.S. military to respond immediately to presidential orders.

The maintenance of this constant state of readiness had an impact on all U.S. military bases, including Fort Leonard Wood. Training and technological advances required advanced preparation for military operations in a multitude of international theaters. As the country was in the midst of budgetary challenges because of growing economic concerns, the military was adjusting to training consolidations. While many military bases and their surrounding communities were being threatened with base closures, Fort Leonard Wood and its surrounding areas would experience one of its greatest periods of growth and expansion.

1980s, Cold War

Base Activities

Activities on the base in the early 1980s included both destruction and construction. In March 1980, many of the World War II era buildings were in disrepair and were stripped of their lumber and usable fixtures. The 5th Engineer Battalion cleaned up the chunks of cement and piles of unusable materials. These were hauled to what was referred to as "The Million Dollar Hole," where the material was used to prevent erosion along a streambed that flowed through the training area.

With just under 15,000 permanent residents on the base, the need for advanced educational opportunities became apparent. Several of the state schools and universities opened satellite campuses on the base at the Truman Education Center. There was a wide range of educational opportunities, from studying conversational German to engineering management. Courses leading to accredited college bachelors and masters degrees were also offered. One soldier said, "The name of the game is initiative. If you want to waste your service career getting drunk at the club or getting high in the barracks, no one's going to stop you." "But," he continued, "if you want to work on getting a degree or just studying something that interests you, the people at Truman Education Center will really help you reach your goal." Many soldiers took advantage of this opportunity to receive a college degree. In 1988, about 2,000 individuals received instruction at the education center. More than 1,100 individuals received tuition assistance to the seven colleges that operated from the Education Center. The center's staff also assisted with the Basic Skills Education Program which had a total enrollment, in 1988, of 883 individuals.

Many college-level teachers from around the state were needed to fill the class needs at the Truman Education Center. One such teacher was Dan Beach from Springfield, Missouri. He taught at the Drury Center at Fort Leonard Wood one night a week for sixteen years, from 1983 until 1999. As a graduate teacher in education, he was instrumental in assisting service men and women who were leaving military service and seeking a career in teaching. Many schools in the Fort Leonard Wood region benefited from the student-teachers in their schools from the Truman Education Center. One of Beach's more memorable teaching experiences was entering his classroom early only to find numerous claymore mines (explosives) scattered about. The previous class instructors had taken a break and had not collected their equipment. Dan waited until the room was clear.

A longtime local resident, Kenny Foster, was born on farmland that became the eastern part of Fort Leonard Wood. He attended the Rolling Heath School, still on base today. He taught in an area school eleven years teaching biology and working as a basketball coach. He worked on the base from 1966-93 as safety officer with responsibilities for accident prevention. This required inspecting training ranges as well as traffic control.

In 1982, the prominence of the base's engineer training was evidenced by the presence of the 4th Training Brigade used to train engineers from fifteen foreign countries. In the summer of 1988, the Foreign Liaison Office transferred to Fort Leonard Wood. These officers provided a vital link between the Engineer School and its counterparts in Canada, Germany and France.

In early 1988, the Engineer Center "was tasked to develop a complete Mine Awareness/Countermine Training Program for United Nations personnel and the people of a country recovering from a Soviet backed counter-insurgency war." Soviet backed forces had planted from five to fifteen million mines in their efforts to counter an insurgent movement. This important center at Fort Leonard Wood would provide "evaluations and advice to the Defense Department, State Department, United Nations Commission for Refugees and host country officials."

New technologies from the antiquated World War II equipment would provide improvement in "doing greater work, have a higher degree of mechanical reliability and save scarce manpower." Much of the equipment engineers used in training dated from the 1940s and 1950s. The basic tools of the Army Engineers were undergoing modernization or replacement. Examples of the improved equipment were the Armored Combat Earthmover (ACE), Counterobstacle Vehicle (COV), Combat Mobility Vehicle (CMV), Combat Gap Crosser (CGC), Combat Excavator (CEX) and Mine Dispensing Vehicle (MDV).

The base has actively hosted the Missouri Special Olympics since 1975. In the summer of 1988, more than 3,000 Olympians, coaches and chaperones attended along with an estimated 4,000 visitors. The troops based at the fort were very supportive and committed in their assistance to this event.

Technology, that is taken for granted today, was innovative in the late 1980s. Use of computers and automation devices was now available to improve office efficiency and mission support enhancement.

To provide awareness of base activities and events, a base newspaper continued its publication. In the earliest days, it was called the *Fort Wood News*. In the late 1950s, it was changed to the *Fort Wood Sentinel*. In January 1966, it was changed to the *Guidon*. In 1988, in honor of the new Engineer Center, it was changed to *Essayons*. With the addition of the Chemical and Military Police schools in the 1990s, it was changed back to its current name, the *Guidon*.

In 1982, the much needed Duvall Maintenance Complex was constructed on the base. New construction on post also included the Engineer Center Complex, which comprised the post headquarters, an academic building and a library structure. New unaccompanied officers quarters for students in the Engineer courses was also a 1989 work project. At the same time, a new reception center, Grant Hall, was completed and occupied by the 43rd AC Battalion.

Training was still the central purpose of Fort Leonard Wood. During the 1980s, an average of 30,000 soldiers received basic and advanced training. An average of nearly 35,000 received training in National Guard and Army Reserves roles. All of the training continued in the midst of new construction on base and the preparation for base expansion.

Even with all of its activities, the post was not exempt from the national effort to reign in the economy. In May 1988, the Department of Army directed a 10-19 percent cut in material development as part of an overall organization and equipment adjustment.

Base Expansion

For over twenty years there had been discussions among military leaders and politicians about having two separate engineer training centers—Fort Belvoir, Virginia and Fort Leonard Wood, Missouri. There was concern over the lack of standardization at both training centers; duplication of some personnel function, equipment and programs; and the expense of moving soldiers from one facility to the other.

On February 28, 1985, authoritative word came from the Department of Army Headquarters in their Public Affairs Office Release No. 22:

> The Department of Army has announced that elements of the U.S. Army Engineer School (USAES), a U. S. Army service school, will be relocated from Fort Belvoir, Virginia, to Fort Leonard Wood, Missouri. Groundbreaking for the first building required by USAES could take place as early as the fall of 1986 and it is anticipated that total construction time will be 2.5 to 3 years. Thus, the personnel influx could begin in late 1989 or early 1990.

The release also reported that:

> The move means a major impact on Fort Leonard Wood and mid-Missouri. Nearly 2,000 additional personnel will be assigned to Fort Leonard Wood. This includes about 800 permanently assigned soldiers, 300 civilian personnel, and 700 commissioned and noncommissioned officers as students in residence at any one time. A total of 3,000 students will be trained annually. These totals do not include family members so the net population gain will, in fact, be even greater.
>
> Fort Leonard Wood's annual expenditures will increase by about $38 million from the present level of more than $462 million. That equates the post expenditures of nearly $1.5 million per day.
>
> Construction costs resulting from this move are estimated at about $40 million. Annual savings will approach $14 million. Thus, the initial cost of the move will be paid for in just over four years.

In March 1985, Commanding General John Moellering had reported that the "urban sprawl" near Fort Belvoir severely restricted the engineer officer's training. The soldiers would have to travel about 60 miles to Fort A.P. Hill for additional ordinance training. Moellering added that, "while Fort Belovoir has no firing ranges at all, we have 31 very high class ranges here at Fort Wood."

The final decision for a move did not go without considerable political opposition. Congressional delegations from Maryland and

Virginia rose in Congress to voice their opposition to the approved move. U.S. Representative Stanford Parris, Republican from Virginia, said, "I thought this snake was killed two years ago. I just think the human impact and sensitivity of these kind of decisions need to be considered more." U.S. Representative Ike Skelton, Democrat from Missouri, countered by arguing that Fort Belvoir's limited 8,500 acres needed to be compared with Fort Wood's 63,000 acres. Some opposition came from what Skelton labeled "old generals, old colonel types, who had gone to the engineer school, didn't want it moved and they just weren't for it. It had always been done that way." Fort Belvoir would change its identity and military mission with the placement there of an intelligence and command facility at an estimated $38 million and completion by 1988.

With the announcement and plans for consolidation of the two engineer training bases, local communities around Fort Leonard Wood foresaw boom days ahead. The impact of the fort on the area was noted in a 1980 study done by a Jefferson City, Missouri, community planning consultant. It was indicated that, "the Waynesville-St. Robert area receives 18 cents on every dollar of the fort's payroll." That number would only increase in percentage and total.

Representative Skelton reported that, "almost $60 million worth of construction, including five school buildings, will be completed before the engineer center opens in 1989." He also indicated that "the Army has no plans to construct additional homes or apartments on the post to offset the influx of 1,000 permanently assigned soldiers and civilians." The fort already housed more than 13,000 military personnel, 60 percent of them trainees, and employed 4,200 civilians. That meant that the local communities would house the new arrivals and families. A big need for housing units off of the base was consequently indicated.

Another report provided information for an employment boom at the fort. Colonel Beardslee, new director of the Engineer Center Transition Office, reported that, "only an estimated 5-8 percent of the civilian employees at Fort Belvoir are expected to relocate to Fort Wood. That means we are going to have to hire them from somewhere." In reality only 71 of the 223 civilians employed at Fort Belvoir chose to move to Missouri (31%). It was also indicated that these civilian employees would earn from $30-50,000 a year. It was estimated that approximately 670 new federal jobs would be added to the installation. Estimated off-post jobs created by the increase would be approximately 900.

There were three major dimensions of the relocation. The first was the design and construction of the new facilities needed to house the school and library. The second involved moving organizations and equipment from Fort Belvoir. The last aspect of the relocation was the moving of people. The consolidation of the engineer education and training, which would be complete at the end of 1989, would mark the first time the engineers have had a single home in 50 years.

Groundbreaking ceremonies were held on August 15, 1987, with military and political dignitaries in attendance. Construction of the $56.5 million project would be under the supervision of the Corps of Engineer's Kansas City District. The J.S. Alberici Construction Company, Inc., of St. Louis was awarded the $28,630,132 contract for the headquarters/administration building and the academic building. However, the construction process did not go without incidents of controversy. Representative Skelton made charges against the Reagan administration's efforts at the "privatization" of the federal government in awarding some construction contracts to private firms. Skelton met with the secretary of the Army to discuss this process. Army officials said the competitive awarding of contracts could save the government $17 million.

Rep. Skelton recommended naming the new headquarters building after General William Hoge, a native of Lexington whose father served as a leader of Wentworth Military Academy. General Hoge was a West Point graduate and a highly decorated battalion commander during WWI. In WWII, he directed the Ninth Armored Division in the "pivotal, bloody, and game-changing capture of the Remagen Bridge over the Rhine River in Germany in March 1945." Hoge Hall is a monument to his military service.

There were other building construction projects underway on the base, including a new Army reception center for the 43rd AG Battalion. This new brick facility would replace the numerous early satellite buildings to consolidate the in-processing functions. A new Army reserve center was finished in early 1987 that replaced and consolidated a variety of meeting and training base locations. A new heavy equipment building, Kawamura Hall, was completed in 1986. In 1985, a new "vertical construction skills" building was completed to

Photographs

Originally appeared in the Fort Leonard Wood 1967 yearbook. Used with permission.

Originally appeared in the Fort Leonard Wood yearbook. Used with permission.

Training at Fort Leonard Wood. Courtesy of the Engineer History Office.

Training at Fort Leonard Wood. Courtesy of the Engineer History Office.

Training at Fort Leonard Wood. Courtesy of the Engineer History Office.

Training at Fort Leonard Wood. Courtesy of the Engineer History Office.

Training at Fort Leonard Wood. Courtesy of the Engineer History Office.

Photographs

Training at Fort Leonard Wood. Courtesy of the Engineer History Office.

Training at Fort Leonard Wood. Courtesy of the Engineer History Office.

Training at Fort Leonard Wood. Courtesy of the Engineer History Office.

Photographs

Training at Fort Leonard Wood. Courtesy of the Engineer History Office.

Training at Fort Leonard Wood. Courtesy of the Engineer History Office.

Training at Fort Leonard Wood. Courtesy of the Engineer History Office.

Training at Fort Leonard Wood. Courtesy of the Engineer History Office.

Photographs

Training at Fort Leonard Wood. Courtesy of the Engineer History Office.

Training at Fort Leonard Wood. Courtesy of the Engineer History Office.

Training at Fort Leonard Wood. Courtesy of the Engineer History Office.

Photographs

Training at Fort Leonard Wood. Courtesy of the Engineer History Office.

Training at Fort Leonard Wood. Courtesy of the Engineer History Office.

Training at Fort Leonard Wood. Courtesy of the Engineer History Office.

Training at Fort Leonard Wood. Courtesy of the Engineer History Office.

Training at Fort Leonard Wood. Courtesy of the Engineer History Office.

Laying a wreath at Fort Leonard Wood. Courtesy of the Engineer History Office.

Originally appeared in the Fort Leonard Wood 1967 yearbook. Used with permission.

Photographs

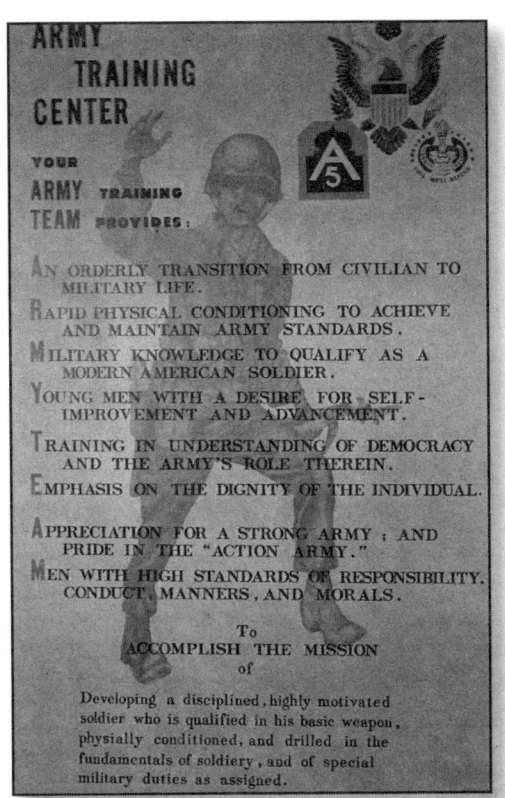

Originally appeared in the Fort Leonard Wood 1967 yearbook. Used with permission.

New base headquarters complex, 1980s. Courtesy of the Engineer History Office.

New base headquarters complex, 1980s. Courtesy of the Engineer History Office.

Photographs

New base headquarters complex, 1980s. Courtesy of the Engineer History Office.

MWR Outdoor Adventure Center (skeet, archery, rental facility and outdoor gear shop), author's photo.

Lieber Heights residential area, Ft. Leonard Wood. Originally appeared in the Fort Leonard Wood 1967 yearbook. Used with permission.

The John B. Mahaffey Museum Complex housing the Engineering, Bio/Chemical and Military Police Museums, author's photo.

Photographs

Nutter Fieldhouse, 1942, public meetings, banquets, international events, author's photo.

Forney Field Airport, with flights to St. Louis, author's photo.

Mahaffey Museum Complex from History Park, featuring a UH-1B "Iroquois" helicopter from the 560th Military Police Company, author's photo.

The History of Fort Leonard Wood, Missouri

Building 1316. This barracks facility, built in 1941, is preserved to show what life was like for basic trainees for decades at the fort, author's photo.

Barracks, Building 1316

Building 1316 is a 63-man enlisted barracks constructed in 1941. It was designed to house 53 enlisted soldiers and 10 non-commissioned officers. In 1944, the single beds were replaced with bunk beds increasing the capacity of the barracks to more than 120 soldiers.

The building is a standard design barracks of the 700-series mobilization construction program. It contained 4,720 square feet including the mechanical room. It was heated by a forced-air furnace which burned coal. The hot water for the latrine was also heated by coal which was stored in bins outside of the building. In 1941, 613 barracks of this type were built on Fort Leonard Wood at a cost of $19,505 per building.

This building is not open to the public at this time.

Sign outside of Building 1316, one of the original barracks constructed in 1941, author's photo.

Photographs

A line of racks in first floor of barracks Building 1316, author's photo.

This close-up shows some of the gear a basic trainee was expected to maintain, author's photo.

General Sherman M4 US Army tank on display at the museum. The M4 was used from 1942 until 1955, author's photo.

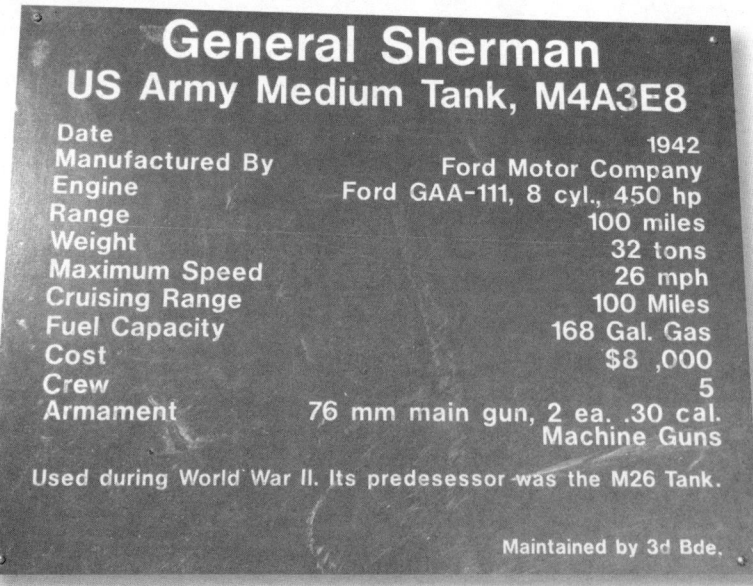

Sign detailing the specifications of the Sherman M4 tank, author's photo.

Main Post Chapel, interdenominational, author's photo.

4th Maneuver Enhancement Brigade Headquarters, author's photo.

Grant Hall, reception center. After 72 hours, recruits are sent to their unit assignments, author's photo.

1st Lt. Terry facility, Initial Response training, author's photo.

Photographs

Truman Education Center (six colleges and universities), author's photo.

Base and Community Hospital, author's photo.

Thurman Hall entrance, Military Police (MP) and Chemical Biological Radiological and Nuclear (CBRN) Schools, author's photo.

Star base for basic training, offices, classrooms, barracks, physical training area in center, author's photo.

Photographs

Star base for basic training, offices, classrooms, barracks, physical trianing area in center, author's photo.

Inside the plaza of Star base facility, author's photo.

Hoge Hall, Command Headquarters, built 1988-90, author's photo.

Hoge Hall, Command Headquarters, built 1988-90, author's photo.

Photographs

Lincoln Hall, Clarke Library, author's photo.

Lincoln Hall, Engineer School, author's photo.

Front of Clarke Library, author's photo.

Lincoln Hall, Engineer School, author's photo.

Marine Headquarters building (Marines of the Ozarks), Engineer, all Department of Defense and truck driving, author's photo.

consolidate carpentry, masonry and other skills previously taught in sixteen other locations on base.

Interservice training also continued at Fort Leonard Wood. In 1985, over 1,000 Marine Corps and Air Force enlisted engineers were trained at the base. More than 18,000 soldiers from National Guard and Army Reserve units, Reserve Officer Training Corps, and individually ready reservists also received training support from the post for weekend and annual training periods. Part of the basic training at Fort Dix, New Jersey, was also moved to Fort Leonard Wood. In 1989, basic combat skill instruction was provided for more than 21,000 soldiers. Advanced individual training would add another 10,000 to the total number of soldiers trained.

Accomplishments

In February 1988, the Army announced a major cutback in personnel. WWII had the highest level of trained troops at 5.98 million. The Korean conflict involved 1.59 million soldiers. The peak of the Vietnam War involved 1.57 million. Current troop levels at the end of the 1980s were at 781,000. The projected cuts would cause the Army troop level to shrink to 772,600. The budget savings were necessary to pay for the increased defense buildup of weaponry by the Reagan administration. James Ambrose, undersecretary of the Army, explained that the decreases in manpower would be "accompanied by changes in tactics, such as robot infantrymen and radio-controlled armored cars and aircraft." This philosophy was not well received by many generals already complaining of an Army "too small to meet worldwide commitments now" and that there were not enough "training bases needed to prepare draftees to reinforce and replace volunteer troops."

On October 1, the name of the base was officially changed to the United States Army Engineer Center and Fort Leonard Wood. The expansion of Fort Leonard Wood in the midst of national military budgetary and manpower cutbacks would create even greater prestige and recognition as a major military training facility. This recognition was not only acknowledged by civilians, but also by the military itself. In 1985 and 1987, Fort Leonard Wood was honored with the Army's Installation of Excellence awards. Inspection teams graded military bases including Fort Sill, Oklahoma; Fort Jackson, South Carolina; Fort Benning, Georgia; Fort Knox, Kentucky; and Fort Bliss, Texas.

Judging criteria focused on "imaginative and innovative management measures taken by the installation to increase productivity and improve the overall quality of life." Fort Leonard Wood was cited for efforts "to enhance customer service, long range planning, people programs, facilities and base support." Colonel Thomas B. Reth, director of engineering and housing, said, "Out of 20 evaluation areas, Fort Wood won seven, making Fort Wood the clear winner." Reth added that the award might help dispel the old notion of "Fort Lost in the Woods."

A 1988 Department of Defense impact study estimated that 4,350 people would move into the Fort Leonard Wood area by the completion of the Army Engineering School relocation in August 1989. Representative Skelton said this figure "includes the anticipated number of military and civilian personnel, as well as their spouses and dependents." Included in this number would be over 500 children of school age to be added to the local school system. Further expansion was reflected in April 1989 with a groundbreaking held for the construction of the Unaccompanied Officers Quarters and the library.

On June 1, 1989, the colors of Fort Belvoir were presented to Fort Leonard Wood. At a special ceremony, Missouri Governor John Ashcroft declared the day as Fort Leonard Wood Day in Missouri. He said, "I join Missouri and all the people of the State of Missouri in support of this vital military community and its contributions toward maintaining a strong and effective defense." Representative Skelton said, "The men and women of Fort Leonard Wood, beginning today, will be and are, the finest military engineering school in the free world. You are the front line of our defense and I compliment you."

The December 1989 opening of the $28.5 million U.S. Army Engineer Center sparked expansion at the Pulaski County Army post, he said. "This is not an ordinary training post," Skelton said. "It is here to stay."

Conclusion

At the end of the 1980s, many other communities around the country with military bases were experiencing the negative challenges of change that faced the Fort Leonard Wood area after WWII. But now the communities around Fort Leonard Wood were experiencing the benefits of the military base growing into a national center for engineer training. The 1990s would offer more prestigious growth for the base with additional national training centers.

Chapter Eight
1990s, War on Terrorism

Introduction

The decade of the 1990s held significant challenges and opportunities for Fort Leonard Wood. Challenges came in the form of growth pains, budget cuts, technological advances and training tactics. Opportunities appeared in the form of the base expansion, national recognitions, specialized training and technological advances. Strains developed on both the U.S. economy and military manpower. An international war on terrorism necessitated new war strategies, employment of new technologies and more efficient training. Fort Leonard Wood was continually challenged with serious budgetary cutbacks in operations, personnel and equipment while the demand for the number of troops being trained increased.

International Crises

International events had a great impact on the U.S. military and Fort Leonard Wood during the 1990s. The fall of the Soviet Union may have led many to assume that the threat of armed conflict had passed. But a new kind of threat from a new kind of enemy would challenge such assumptions. The major event of the decade was the invasion of Kuwait by Iraq, leading to major American military participation in Operations Desert Shield and Desert Storm. U.S. military forces were also involved in smaller scale operations in Zaire, Sierra Leone, Somalia, Macedonia, Haiti, Bosnia, Liberia, Central African Republic, Albania, Congo, Gabon, Cambodia, Kenya, Tanzania, East Timor and Serbia. Many of these international operations were to assist in humanitarian needs and protection of specific national interests.

Base Activities

The 1990s saw one of the most dramatic changes and challenges

1990s, War on Terrorism

in activities on Fort Leonard Wood. If the 1980s, with the arrival of the Fort Belvoir Engineer School consolidation, saw change, it was indeed minor in comparison to the events of 1990. All areas of the base would be affected by funding cuts, increased training demands, initiation of gender-integrated training, employment of interservice training, demands for modernization of facilities and equipment, increased international deployments and the impact of the significant base expansion through the Base Closure and Realignment Commission.

Funding

National and international economic demands caused consolidation and reorganization of the base's traditional structures. Decreased federal funding challenged the base to evaluate expenditures and personnel needs. Increased expenditures for deployments in the conflicts in Iraq affected the base's use of manpower and resources. Even after two years of initial conflict, there were continued residual expenses involving storage of materials, equipment repair and outprocessing of personnel. In addition, brief government shutdowns caused disruption of the civilian workforce and operations. In spite of the increased expenses, military funding for the installation continued to be decreased throughout most of the 1990s. 1997 was the only year when the base received an actual 10% funding increase.

To offset the operating fund decreases, the base discovered numerous ways of cutting costs and increasing work efficiency. In 1990, Fort Leonard Wood was challenged to reduce expenditures by $11.59 million. A Systematic Productivity Improvement Review reported 66 initiatives saving $23.6 million, or 203 percent of the established goal. The use of electronic processing of forms, new long distance phone service, contractual disposition of waste paper, and the new Defense Reutilization and Marketing on-line service, were all means to achieve efficiency and savings.

The base also benefited financially through several Army actions. In 1992, Fort Leonard Wood was designated to be a closeout installation, receiving excess funds from other installations being closed out. With the Interservice Training Review Organization training for Air Force, Navy and Marine Corps personnel, the base received reimbursement for those training expenses from the other military branches.

Training

While the traditional training programs of the base continued, additional training demands would strain leadership and resources. The annual training load at Fort Leonard Wood was divided into basic training, advanced training, enlisted engineer training, noncommissioned officers, warrant and commissioned officers. These totaled just under 30,000 troops. In addition, training also included reserve units and National Guard units. Referred to as the "summer surge," troop levels trained could potentially equal 30,000. Reserve personnel and National Guard units were called into active duty by the president in response to the Iraqi invasion of Kuwait in 1990. Fort Leonard Wood was selected to mobilize these troops.

Between September 1 and December 31, sixteen United States Army Reserve and National Guard units were mobilized at the base. It was discovered that well over half of the reservists could not pass the Army Physical Fitness Tests (APFT). In one unit, only 23 of the 170 individuals tested passed the APFT.

In the early 1990s, there was an attempt by several congressmen and a Training and Doctrine Command (TRADOC) study team to eliminate the Basic Combat Training (BCT) at Fort Leonard Wood. After further study, TRADOC announced the decision to discontinue any effort to eliminate BCT at any of the four existing sites.

In addition to the training of troops, the base conducted numerous training sessions for civilian employees. These training sessions included computer literacy and software classes. Other classes offered specific skill enhancement training.

Several training assignments were given to Fort Leonard Wood. In May, the post received a request to train Haitian national Police cadets. The training was coordinated through the Departments of State, Justice and Defense.

Approximately 4,000 cadets were trained at a cost of $4 million. There was also an increased emphasis on Operations Other Than War (OOTW). These training activities were associated with humanitarian and specialized operations, which required substantial study and changes in the areas of concept and doctrine. Training exercises included responding to an explosion by a terrorist group, medical response to an earthquake in the New Madrid area of Southeast Missouri, and anti-personnel landmines. In 1998, the Countermine Test Facility was completed. Non-resident training began with the use of Distance Learning, Engineer School homepage and the Total Army School System, which offered support to off-post organizations and units.

Gender Integration

One of the major changes in 1991 was the reinstitution of basic training for female soldiers at Fort Leonard Wood. The installation had trained women in the late '70s and early '80s, but that was soon eliminated. With the arrival of female trainees, there were some adjustments. Most of these dealt with facilities requirements such as clothing and equipment issue. In June, the post received its first all-female BCT in ten years.

In 1994, the Secretary of the Army announced new guidelines for women in combat. This led to the opening of a number of positions in the combat engineer companies and platoons.

Fort Sill, in Oklahoma, was the only other gender-integrated Army base in the United States. Several challenges arose immediately upon implementation of the new program. A goal was established to maintain 25-40 percent female-to-male ratios in basic training. This produced reorganization of housing needs. It was discovered that arriving young women were in poor physical condition, which required training adjustments and standards. Drill Sergeants had not been trained in gender-integrated issues.

During the first year, charges of sexual harassment at Army installations led to an Army-wide hotline. Twenty-three of these calls involved Fort Leonard Wood's 3rd Brigade. The Army responded with increased sexual harassment classes and instructor training. However, the annual reports of sexual harassment continue even up to the present day.

Interservice Training

The effort to provide more efficient national specialty training in engineering, military police, chemical warfare and transportation, created an interservice expansion of training at Fort Leonard Wood. In 1994, the installation became a joint services training center with the approval of recommendations of the Interservice Training Review Organization (ITRO). Fort Leonard Wood was becoming "purple"—a joint services installation.

Specialized training was now being provided for the Air Force,

Navy and Marine Corps. Even the reservists of the military branches across the United States would be trained at Fort Leonard Wood.

The base saw an annual increase of approximately 10,000 additional soldiers receiving training in engineering, chemical warfare and motor transport operation. The increase of training requirements brought about increased expenditures of personnel and funding.

Increases in housing, equipment and training facilities brought new operating challenges for the base. To address these new expenditures, an Interservice Support Agreement was established to provide fair reimbursement from the budgets of the military branches to Fort Leonard Wood.

The base also continued to be called on for the training of specialized, international military officers. TRADOC now looked to Fort Leonard Wood leadership in two areas: the development of support agreements for foreign liaison offices and the resolution of issues involving interservice training under Interservice Training Review Organization (ITRO).

Facilities/Equipment

With the requirements for additional training, Fort Leonard Wood faced decisions regarding whether to renovate existing facilities or demolish existing structures to make way for new construction. It was also determined that "25 percent of the base's automation equipment had exceeded its life cycle." Other problems to be addressed were "barracks without air conditioning, a shortage of computers for company personnel to use and inadequately stocked supply rooms." In the midst of dwindling annual operating budgets, funding for renovations, new construction and new equipment required substantial documentation and support.

Designed decades earlier, the barracks reflected a training system and military population unlike that of the 1990s. Another immediate concern was the renovation of Army Family Housing. Other areas of upgrading existing facilities included "vehicle exhaust system improvements, replacement of the physical security systems on base, bringing the internet to the installation, changes in the post's telephone system, warehouse space and a reutilization program."

Throughout the 1990s, Fort Leonard Wood was able to secure funding for significant new construction projects. Among these were a new youth activities center, a new commissary, a new instructional

1990s, War on Terrorism

media center, a field sports complex, a unit guest house, a physical fitness center, soldier service center, a new child development center, the Wallace Swimming Pool and a launching ramp at the Lake of the Ozarks. Renovation projects included expansion of the post exchange and officers club. Providing organizational efficiency on base were an Energy Monitoring and Control System and a Life Cycle Replacement program for new computer equipment.

Deployments

With Fort Leonard Wood assuming greater responsibility for basic and specialized training of troops, the demands for deployment of these troops for national and international needs greatly increased. The two major deployments for the base involved participation in Operations Desert Shield and Desert Storm.

Almost 4,000 soldiers passed through the installation as part of the operations in the Persian Gulf. Three Fort Leonard Wood units, the 5th Engineer Battalion, 93rd Evacuation Hospital and the 515th Engineer Company (Pipeline) served in the Persian Gulf. Sixteen Army Reserve units, including more than 1,300 soldiers, were processed at Fort Leonard Wood. Nine National Guard units, with almost 1,000 soldiers, were also processed on the base for combat action. More than 1,500 Individual Ready Reserve members were also involved, along with 51 retired military personnel and 15 Individual Mobilization Augmentees. Each arriving unit went through a form of initial entry training in preparation for deployment. Of twelve units evaluated for physical fitness, only one had more than 50 percent of its members pass the APFT. One unit was unable to qualify more than 30 percent of its personnel, even after three tries.

Getting deploying troops from central Missouri to an Airport or Seaport of Embarkation was a major task. It was reported that:

> The transportation staff coordinated the loading of 348 railcars, 26 military aircraft and 14 charter aircraft. Personnel movement was generally by air; some equipment also moved by air, but other materials went by sea. Significantly, there were no late arrivals to Saudi Arabia or missed port calls in spite of the 1,900 tons of equipment and cargo and approximately 2,400 personnel were moved. Supply staff were able to

acquire 2,200 sets of chemical protective garments for mobilizing units from European war stocks.

Virtually every area on post participated in some way in advancing the defense of Saudi Arabia and the liberation of Kuwait. "Not since the Vietnam War had Fort Leonard Wood been so involved in combat, combat support and combat service support of Active and Reserve Component forces. Fully 50 percent of the enlisted strength of the total engineer force deployed to SW Asia." Brigadier General Joe E. Ballard observed:

> In that (Desert Shield and Desert Storm), we faced a threat which our army had not addressed for almost fifty years. The Engineer Center contributed to the success in the Persian Gulf through solutions to complex breaching problems, mobile training teams, equipment modifications and the mobilization and deployment of active and Reserve Component units and personnel.

In the 1990s, after the Persian Gulf deployments, Fort Leonard Wood's specialized training was in demand in other international locations. In 1998, the post received taskings for more than 1,100 personnel for service in the United States and overseas. This was a 154 percent increase in requirements fielded by the installation. A mobile training team was sent to Europe to work with units preparing for movement to Southwest Asia. Military Police personnel were assigned to Honduras and then to Panama in support of operation "Promote Liberty." Other personnel were deployed to Cuba, Korea, Italy, Ecuador, Nicaragua, Chad, Jordan, Mauritania, Somalia for Operation Restore Hope, and to Egypt for Operation Bright Star. The Countermine Training Support Center conducted training for humanitarian demining operations in Chad, Ethiopia, Bosnia and six other countries. The Military Working Dog section sent several training teams to Puerto Rico and Williamsburg, Virginia.

Other Activities

There were other activities involving the base directly and indirectly in the 1990s. Locally, there was an ongoing process of ensuring that local business establishments near the base adhered to state requirements. On April 26, 1991, the Missouri Division of Liquor Control announced that eight St. Robert establishments had their liquor licenses suspended. Most violations included "failure to report an illegal or violent act; failure to cooperate, sale or supply to an intoxicated person; operating a disorderly place, improper acts and sale, supply or consumption by a minor."

Of significant local, state and national interest was the legal case of Dr. and Reserve Captain Yolanda Huet-Vaughn, a physician from Kansas City. When her reserve medical unit was sent to serve in Saudi Arabia, she refused to accompany her unit. A seven-member military jury in August 1991, handed down a 30-month prison sentence. Upon appeal to Fort Leonard Wood commanding general, Major General Daniel W. Christman, her defense attorney, Luther West of Baltimore, argued that she opposed the war on moral grounds but had not left her unit to avoid hazardous duty. Assistant Prosecutor David Harney argued that, "The sick and wounded needed her medical skills, not her opinions." After arguments, General Christman ruled that her 30-month sentence be reduced to 15 months.

On the base, there were many new activities occurring in the 1990s. At the beginning of the 1990s, the Secretary of Defense announced a department-wide hiring freeze. The impact on Fort Leonard Wood was to reduce the employees on base from 1,458 to 1,374. An annual occurrence on base since the 1970s was known as the "Exodus." Holiday leave was offered "to soldiers in basic training, advanced training, officer basic and advanced courses and non-commissioned officer courses." About 4,000 to 5,000 soldiers participated in this two-week leave program. But events in the Persian Gulf required the cancellation of the program in 1991. As a result, many families remained on base. The local community rallied in support of the troops by offering special activities, gifts and holiday celebrations for those unable to leave the base.

The new U.S. Army Engineer Museum had a grand opening on May 10. Classes for English as a second language were offered at the Truman Education Building. Beginning classes had students from 11 countries and attracted some of the international military students' wives. The beginning classes averaged 15 to 16 per class. A number of family support services and activities were also conducted on the base. Among these were Family Day, where the 87th Engineers were able to introduce their spouses and children to their workplace; observation of Child Abuse and Neglect Prevention Month, with a visit from Spiderman; and World Health Day. Other special activities included the opening of a recycling center on Earth Day, 1990, with the center collecting "approximately 54 tons of household items and approximately 165 tons of office paper."

A special office was opened by the Staff Judge Advocate's Office devoted to income tax preparation. In the first three months of operation personnel assisted more than 12,000 individuals with their taxes. Special events on base were the annual Missouri Special Olympics, a reunion of twenty former German prisoners of war and the installation's 50th anniversary celebration.

Statistics from the mid-1990s were given to provide a picture of the military base:

> Post Average Daytime Population: FY 95, 18,921; FY 96, 21,113
> Facilities: Recreational, 72; Ranges, 39; Training, 111; Family Quarters, 2,864
> Land Area: Total on Post, 62,911 acres
> College Enrollment on Post: 10,345
> Religious Service Attendance: 200,000+

Virgie Mahan provides a more personal account of the base and the community. She served as the "temporary" Public Affairs Office director from 1983 until 1996. She fought the tradition of no women having official leadership roles on base. She served as the liaison with the local Committee of Fifty in promoting the base in the community. After retiring from the base, she eventually served as chairman of the Committee of Fifty from 1998-2011. She works with military internationals and the community. She stated that, "Fort Leonard Wood brought diversity to mid-Missouri through race and women's rights."

Base Expansions

The 1990s saw new directions and directives for the United States Army. Terrorist acts against American embassies in Africa raised concerns about the security of all military installations. Questions of budget and training logistics renewed interest in examining the structure and operations of the military. A new operational environment was established with two main ideas. "First, the Army would respond to challenges and deploy from bases in the continental United States. This was a change from the 'Cold War' philosophy of overseas forces being first responders. Second, technology had transformed both the 'threat' and the means to answer the threat." As a result, new initiatives from command moved to create a "force more mobile and capable than the existing service... The solution was to produce a medium-weight organization with greater survivability and combat power than the light division and greater strategic mobility than the heavy division."

BRAC (Base Realignment and Closure)

As with past national, military employments, the cessation of combat activities and increased budgetary demands required a serious exercise in the downsizing of the U. S. military and significant economic savings in the process. Serious debate was underway in the department of defense regarding the use of military force through size and commitment. Nontraditional warfare called for renewed evaluation of training procedures and the employment of new technological advances in military operations.

In February 1991, Secretary of Defense Dick Cheney presented a recommendation of closing 43 military bases as well as a 25 percent reduction in the military force by 1995. He stated "smaller forces need fewer bases." The list of recommended base closings was politically volatile and Congress had great desire to avoid any decision making in the process. A Base Realignment and Closure Commission was established to review the list and make any adjustments and present it to President Bush. Neither the president nor Congress could make any changes in the list. It had to be accepted or rejected in its entirety within 45 days.

When the expected cry of opposition occurred, Cheney replied, "When you've got something desirable in your back yard, then it becomes the NOOMBY (Not Out Of My Backyard) syndrome—keep around, regardless of other considerations." One of the more controversial closures involved Fort McClellan in Anniston, Alabama. The Alabama congressional delegation was powerful and experienced. In 1993, Missouri U. S. Representative Ike Skelton (Democrat) reported that "of the Army's top 13 training facilities, Fort Leonard Wood was ranked fifth in significance to national security; Fort McClellan was ranked ninth." Fort McClellan housed the biological-chemical and military police training centers. Alabama warned Missouri of the dangers and difficulties of trying to secure the proposed transfer of the bio-chemical warfare training to Missouri. When congressional floor debates pitted Alabama against Missouri, Skelton proclaimed effectively, "The Alabama people are saying, 'Missouri, you don't want this, it's dangerous.' Well, if it's so dangerous, why do they want to keep it?"

On May 26, 1991, a key commission vote was taken on the future of Fort Leonard Wood. By a very narrow 4-3 vote, Fort Leonard Wood was kept off of the potential base closure list. "Fort Leonard Wood dodged a silver bullet this morning," said U. S. Representative Ike Skelton, then a senior member of the House Armed Services Committee who represented Fort Leonard Wood.

As expected, the political debates and activities in Congress on issues of base realignments and closures were fierce. Local economies and development were at stake. Senator Howell Heflin, Democrat from Tuscumbia, Alabama said, "We're greatly disappointed. We're going to do everything we can to keep this disastrous decision from going into effect." A contentious debate in the commission deadlocked with a 4-4 vote to keep Fort McClellen open. Only after a compromise "to keep the chemical school and its one-of-a-kind training facility at the Alabama base until all the necessary facilities are built at Fort Leonard Wood," did the commission finally agree by an 8-0 vote to agree to most of the Army's plans for realignment and closures.

In review of the process, Missouri Republican Senator John Ashcroft noted that:

> Alabama used all kinds of tactics to save Fort McClellan, including amendments slipped into legislation on the Senate floor and a series of letters raising numerous questions about the move. Despite a series of smokescreens that drifted from Alabama, I'm pleased that the majority of the BRAC members saw through the fog and critically analyzed the relevant issues.

The combination of political skill from the Missouri congressional delegation and the growing reputation of Fort Leonard Wood's ability to effectively handle base expansion would continue to ensure a solid future for the base and surrounding communities.

April 1991, saw the return of troops from Desert Storm, including reservists and national guardsmen, with a display of welcoming parades and ceremonies from local communities. The national projected base realignment and closures were estimated to cost $5.7 billion between 1992 and 1997. The savings in that period would amount to $6.5 billion—a gain of $850 million. *USA Today* reported that "Under the plan, Fort Leonard Wood is one of a handful of installations that would grow when units from those being closed are consolidated." Local news reports that month indicated that the decisions of base closures would benefit Fort Leonard Wood and the surrounding areas. Chris Morrisey, executive director of the Waynesville/St. Robert Chamber of Commerce said that "the addition of the estimated 6,000 personnel could have a $100 million impact on the regional community."

In October 1991, Fort Leonard Wood Public Affairs Officer Brad Rose reported that three projects costing $11 million were due to be completed at the end of the year, and four more projects were scheduled to begin within a month. These projects included renovations to the post exchange, a 70-unit guest house, sixteen athletic fields for sports, an instructional media center, an automated M-16 qualifications range, expansion of the General Wood Army Community Hospital and a physical fitness center.

U.S. Representative Ike Skelton, serving on the House Armed Services Committee, said he was awaiting authorization from Congress for more than $15 million in construction at the post during the next year. He anticipated the appropriations budget for a $12 million administrative building and a $3 million child development center would be approved soon.

In 1995, President Clinton appointed S. Lee Kling to the Base Realignment and Closure Commission. Kling was a successful banker and insurance executive from St. Louis. The commission had visited over seventy military installations and held multiple hearings across the nation before announcing their decisions. Representative Skelton acknowledges Kling's importance and credits him with the actual expansion of Fort Leonard Wood. Skelton said, "I give Lee Kling enormous credit for his relentless advocacy of Missouri and Fort Wood. Lee passed away in 2008 at age seventy-nine, and Missouri lost a great statesman and I lost a great friend. But his legacy in support of Fort Leonard Wood and his home state is substantial and should never be forgotten."

The complete transfer of the Chemical and Military Police schools from Fort McClellan to Fort Leonard Wood occurred in 1999. It took about seven years and cost $200 million.

With the official opening of the Chemical and Military Police schools and the presence of the engineer training center, the Maneuver Support Center (MANSCEN) was formed. A weeklong celebration was held on the base in the first week of October. Thousands of people attended the events "with dignitaries and distinguished guests from the nation's capitol and military installations around the world." Later that same month, the first soldiers used live agents at the Fort Leonard Wood Chemical Defense Training Facility (CDTF).

Other actions of BRAC added to the growing responsibilities placed on the base. Because of the closing of bases in the central United States, the Casualty Assistance Command of the base absorbed the State of Wisconsin, 15 counties in upper Michigan and 61 counties in Illinois. This expansion of responsibilities resulted in an additional workload of 131 casualty cases and 732 line-of-duty cases in 1993.

Additional cuts in the U.S. Military occurred under the Clinton administration's first year. A review by Secretary of Defense Lew Aspen resulted in major cutbacks in manpower. Again, the U.S. Army experienced the deepest cuts. From a peak of 760,000 soldiers in 1991, troop size was reduced to 480,000 in 1998. Justifications were made with the acquisition of "lethal new technologies such as global positioning systems and laser-guided bombs. During the 1990s, new force structure designs called for smaller combat units with more speed and flexibility."

Chemical and Military Police Schools

Major General Robert B. Flowers observed that, "the leadership and staff... moved Fort Leonard Wood into a new era as a major integrating center for the development of doctrine, training, leaders, organization, material and soldiers... it transformed the installation into a multi-discipline, multi-service, center of excellence." While that was indeed a great honor, the transfer of the two schools to Fort Leonard Wood caused some immediate problems. The arriving units needed barracks and administrative spaces for their personnel. Their families needed housing. There would be an increase in the overall demand for soldier and family support.

With the establishment of the MANSCEN, questions of organization needed to be settled. How would the three schools coexist? TRADOC Commander, General William Hartzog, rejected the idea of having three independent schools on the post. "He wanted to achieve as much integration of functions and, hopefully, economy as possible. Consolidation was to be the rule; collocation was the exception."

Transportation Training

By the end of 1993, the Joint Chiefs of Staff decided that there were to be savings by combining motor transport operator training at Fort Leonard Wood. It was agreed that all engineer operator training for the four services would come together on base by October 30, 1995. In the first year, almost 6,000 soldiers graduated from the battalion's motor transport operator course.

Technological Advances

The Iraqi wars saw great advances and operations of new military technology. These had an impact on the training at Fort Leonard Wood. Developments in strategic weaponry of significance were the Lockheed F-117 Nighthawk, with its stealth capabilities, and the Tomahawk cruise missile. The F-117, which flew only two percent of the attack sorties, struck nearly forty percent of the strategic targets. The Tomahawk was a computer-guided missile fired from U.S. combat vessels striking within a few feet of its target through its digital photo imaging. Major General Judd Blaisdell, then director of space operations and integration for the Air Force, noted in evaluations that, "We are getting better and better at bringing the synergies of air, space, and ground forces together."

Other advances in ground force protection were in the form of potential chemical-biological attacks, environmental adaptations and communication. Improved night vision technologies allowed for effective combat performance. Communication from command headquarters to the ground forces created more effective combat activity and protection.

New satellite technologies allowed for the greatest advances in media war coverage since the Vietnam era. Viewing audiences saw the effectiveness of the "smart bombs" and real time combat scenes as never before. The major U.S. television network coverage was enhanced by the emerging 24-hour coverage of CNN. Not only did the American public visually participate in the war actions, but tapes of combat situations provided invaluable opportunities for evaluation and future training of military forces at bases such as Fort Leonard Wood.

Specifically, the base saw vast improvements in training and technology. New computer programs enhanced the training process. Fort Leonard Wood established a number of classrooms geared to computer-based instruction. JANUS is a two-sided war game where players interact with each other and the computer. In 1996 a Maneuver Support Battle Lab was established to analyze combat operations. The Army's first digital division worked with new tools for mapping, charting and imaging. The tools were also in sync with British, Canadian and Australian forces. The Engineer School was also linked to School of Defense Simulation Internet. Improvements were also made in body armor. In addition, improvements were made in the area of Robotics-Technology Insertion Activity.

Accomplishments

Accomplishments at Fort Leonard Wood in the 1990s were numerous. They contributed directly to the victorious combat operations in Iraq. They were able to incorporate further realignments on the base with new and updated facilities. In addition, they were able to expand to interservice training in engineering, biological-chemical warfare, military police and transport operations. They were also able to increase training and operational efficiency with fewer resources available.

In Operation Desert Storm, Fort Leonard Wood contributed 1,400 members of the base's staff for combat operations. Over 4,000 Reserve Component soldiers mobilized, including 16 United States Army Reserve and 9 National Guard units, for action in Iraq.

Major General Daniel Schroeder said that Fort Leonard Wood made a three-fold contribution to Operations Desert Shield/Desert Storm:

> Fort Leonard Wood is a complex organization, today. It is a Training Center. It is a garrison and the

1990s, War on Terrorism

Engineers school. From the training center aspect Fort Leonard Wood provided a force of soldiers trained to standards. Exceptional efforts were put on the garrison side of the facility. Fort Leonard Wood supported the deployments of about 22 units containing more than 3,000 people. The school house side started in September, with the Engineer battalions providing mobile training teams for pre-deployment training.

In Viet Nam we were basically a draft army. The force you saw in the field today in Desert Shield/Desert Storm, you called it a volunteer force, but I would tell you it is a professional force.

Adapting well to the expansion of the engineer training center, the base used the lessons learned to prepare for the expansion on the base of the chemical and military police schools. Additionally, the base also grew to become the national military center for transport operation training. With all of these now established national military centers, the base was also able to incorporate the added responsibility of interservice training for the Air Force, Navy, and Marines Corps.

During 1994, the first gender-integrated training of military forces took place at Fort Leonard Wood. The ensuing challenges of such an operation were met with continued evaluation and improvement.

Perhaps the greatest accomplishment for Fort Leonard Wood in the 1990s was to incorporate all of these additional responsibilities while improving training and operational efficiency using fewer allocated resources.

Annually, TRADOC selects the installation with the best FAO. In 1992, TRADOC awarded Fort Leonard Wood with the best (Functional Area Assessment) FAA for a medium sized TRADOC installation. Criteria for selection is based on installation size and standard of performance in six specific areas of operations; military pay timeliness, discounts lost, prompt payment act (PPA), interest payments, travel work in place (TIP), turnaround time, and civilian pay timeliness. Fort Wood won over Forts Eustis, Gordon, Jackson, Lee and Rucker.

In August, the Mine Detection Facility Working Group visited the installation. Fort Leonard Wood was one of six sites under consideration. The others were Camp A. P. Hill, Eglin Air Force Base, Blossom Point, Maryland, Aberdeen Proving Grounds, and the Naval Air Warfare Center. After all sites were assessed, the working group briefed the Joint Working Group. In early September, Fort Leonard Wood was selected as the preferred location. Official notice came in late November.

Conclusion

The 1990s challenges to the United States military were significant. A new philosophy of a smart war was established with more technological advances and the assumed necessity for fewer troops on the ground. The use of national guard troops for international combat and the multiple, extended tours of all military active duty, placed unique strains on the troops and their families. The recruitment of replacement troops and retention of trained leadership were also challenges.

The 1990s were evolving into a new era for Fort Leonard Wood. The changes and challenges on the base never distracted from the main purpose.

> Fort Leonard Wood's people are the reason we can achieve these goals. The joint team—soldiers, sailors, airmen, marines, civilians and family members— allows us to achieve our objectives, and it is they who will take us into the twenty-first century as a first-line member of America's Defense Team. One Team— One Mission—One Fight! (Major General Robert B. Flowers)

The center refined its gender-integrated training approach, initiated training under the Interservice Training Review Organization (ITRO), and began to lay the foundations for the Maneuver Support Center (MANSCEN) with the arrival of the military police and chemical school. Training and concept developers used computer-based simulations and modeling to enhance training and development system and force structure for the next century.

As the 1990s came to a close, the U. S. military and the expanded operations at Fort Leonard Wood were only beginning the war on terrorism. The 2000s would begin with a new administration and an attack on the homeland that would test the military decisions and expenditures.

Chapter Nine
2000s, WAR ON TERRORISM

Introduction

The national and international events of the 2000s decade would drastically alter the concept of warfare, training and technology for all U.S. military forces. It would also produce national and international debate on the use of weaponry and prisoner interrogation techniques. Questions would also be raised about national and military leadership and the issues of security and personal freedoms.

International Crises

For the U.S. military, the 2000s began much the way the 1990s ended. There were small numbers of military forces deployed for training and assistance to distant international places such as Sierra Leone, Nigeria, Yemen and East Timor. The situation for the entire international community, however, was anything but usual after September 11, 2001. Hijacked airliners crashed into the World Trade Center Towers and the Pentagon that morning which resulted in the deaths of hundreds of Americans and international citizens. Another hijacked air flight was heroically crashed in Pennsylvania before their cowardly attack could be completed. The massive attack and destruction in New York City and Washington, D.C. created universal panic ultimately resulting with unified retaliation for the perpetrators of the cowardly crimes.

On October 7, 2001, U.S. military forces invaded Afghanistan, with Operation Enduring Freedom, in response to the 9/11 attacks. Combat operations began with strategic strikes against Al Qaeda terrorists and their Taliban supporters. Further major action was taken in the war on Iraq through Operation Iraqi Freedom. A limited coalition of national forces invaded Iraq with the goal "to disarm Iraq in pursuit of peace, stability, and security both in the Gulf Region and in the United States."

2000s, War on Terrorism

In addition, continued use of a limited number of U.S. military forces would be deployed during the decade to Yemen, the Philippines, Cote d'Ivoire, Liberia, Georgia, Djibouti, Haiti, Lebanon, Ethiopia, Somalia and Eritrea. However, traditional warfare was assisted by use of smarter training and technology. This applied to the military training facilities such as Fort Leonard Wood.

National Crises

Following the attacks of September 11, the byword for the early 2000s was homeland security. This included greater public awareness, transportation scrutiny and also military installation security. National alert levels were created and color-coded to express the degree of concern of terrorist activities and threats. Air travel, already complicated, was made more difficult with passenger and baggage screening procedures being intensified. Challenges to the public's personal freedoms were made through newly-approved screening of individual communications. Some communities in the United States reacted in anger and fear of Middle Eastern citizens with Muslim/Islamic affiliations.

The direct attack on the Pentagon prompted great concern for the security and safety on the nation's military installations, including Fort Leonard Wood. The base was immediately put on "force protection condition delta," the highest state of alert. A 100 percent check of incoming traffic and personnel created long lines with backedup vehicles. There was increased protection of gathering spots on post, such as the theater, airport and other perceived high-risk targets. Even with this alert level, it was determined that training and graduation would continue as normal. In the immediate following weeks, additional securities were assigned to key buildings, entryways and computer security. Civilian workers were praised for their willingness to serve extra hours during the round-the-clock planning operations at the base's Emergency Operations center. Off-post businesses donated goods and services to show their support for the military.

Major General Anders Aadland, Maneuver Support Center and Fort Leonard Wood Commander, in a speech given to the NCO Academy on base, stated the post's mission:

> It's a mission that will take determination and focus…
> This war is far different from any war the U.S. military

has fought in the past, but the mission at Fort Leonard Wood remains unchanged. Our business here at Fort Leonard Wood is to train soldiers, Marines, airmen and sailors every day to standard. We must continue to feed the Department of Defense those outstanding, young entry troops... The Chemical, Engineer and Military Police schools will play vital roles in the fight against terrorism. The three branches here happen to be absolutely critical to the mission of homeland security. They're all here, so we're going to have to continue to deliver that expertise and that capability. That is absolutely critical.

Base Activities

Fort Leonard Wood was still adjusting to the base expansions of the 1980s and the 1990s. It was evolving as an interservice, gender-integrated military installation. Training and technology would be geared toward the goal of smaller and smarter armed forces, with increased training regarding mobility and computer-aided warfare.

Furthermore, in recognition of the need for constant evaluation of the base's organization and operations, new programs were established to oversee operations in light of national military directives and concerns. An Installation Management Agency (IMA) was formed in October 2002.

It was during this time that the issue of gays in the military was addressed by the Clinton administration's new practice of "don't ask, don't tell." The impact at Fort Leonard Wood led to the dismissal of sixty soldiers in 2005, an increase of 50 percent from 2004.

Construction on base was minimal but significant. In April 2005, a much-needed, new Post Exchange was opened for base personnel and the growing number of retired veterans living in the surrounding communities. In 2008, the Maneuver Enhancement Brigade was activated on the base. In 2009, the United States Army Reserve Division was re-activated on base. In addition, Fort Leonard Wood became the home of the Department of Defense Humanitarian Demining Training Center.

Training continued to be the primary emphasis on the base. On October 1, 2009, Fort Leonard Wood was designated as a Maneuver Support Center of Excellence, one of only eight centers in the United

States. At this time, trainees for military and civilians totaled 80,000 to 90,000 each year. Of the women being trained in the U.S. Army, 42 percent started their careers at Fort Leonard Wood. Nearly 13,800 people from 120 reserve units were now trained on the base each year.

Technological Advances

The advances in technology for training and warfare had grown to very sophisticated levels. There were improvements in reconnaissance with automated drones that could also be armed. Battle operations and progress could be seen in real time video by commanders behind the lines. Such video footage was now a regular feature of military briefings during combat operations. Video technology was also an important part of the training now on most military installations.

Accomplishments

The greatest accomplishment at Fort Leonard Wood during the 2000s was the adjustment of growth in personnel and training responsibilities. There was also adjustment to the reorganization in base structures as required by the new directions from high command. The base accomplished all of these with the usual challenge of decreases in annual funding.

The base museum was expanded to include a wing for the Chemical/Biological School and the Military Police school, in addition to an expanded Engineer school display. Updating of computer equipment and digital training innovations better prepared soldiers for combat situations. Continued improvement in base housing and recreational facilities was well received by the troops and their families. Continuing education classes grew in number, attendance and graduates.

Conclusion

The 2000s would test the new concepts of a smaller and smarter United States Army. While the policies were set at the national level, the development and implementation of these new concepts were accomplished at military installations across the country. In the light of these changes, Fort Leonard Wood had become the premiere Army base in the country. With a unique position as an interservice and gender-integrated center of training, with three schools of operation, its place in national military importance had grown.

National economic pressures would continue to challenge all

military installations; Fort Leonard Wood was much more secure as an ongoing training facility. There would be some minor cutbacks in personnel announced, but the base itself remained sure of its continued role.

Chapter Ten
2010s, War on Terrorism

Introduction

The onset of the present decade of the 2010s would signal a solidifying of the importance and identity of Fort Leonard Wood. The concepts of training would continue to evolve with the development of technological advances. The withdrawal of American combat troops in Iraq and eventually Afghanistan would have an impact on the funding for military operations. That decrease in funding would filter down to the military installations throughout the United States.

International Crises

The U. S. military saw continued service in international arenas. The two major theaters of operation were in Iraq and Afghanistan. In early 2010 in Iraq, "Operation Iraqi Freedom" was replaced with "Operation New Dawn." The training of Iraqi security forces coincided with the reduction of 50,000 American forces. The military presence of U.S. forces continued in Afghanistan. The training of Afghan security forces paved the way for withdrawal of most American combat forces.

In addition to the two major military operations, the U.S. also provided limited, military assistance in a number of other countries including Libya, Somalia, Uganda, Jordan, Turkey, Chad and Mali. There was an increase in the use of drone attacks on discovered terrorist leaders and camps in Somalia, Yemen, Libya Afghanistan and Pakistan. A significant military action was "Operation Neptune Spear" in 2011. A raid in Pakistan by elite American forces killed and captured Osama Bin Laden, designated "Geronimo."

Additional crises occurring in Lybia, Syria and Egypt caused American forces to prepare for potential, limited military action. That preparation took place on U.S. military bases, including Fort Leonard Wood.

National Concerns

National political gridlock continues to grow as a serious concern for the daily operations of military installations. The battle between the executive and the legislative branches is being fought in the arena of budgetary appropriations. A new term, "sequestration," was introduced into the national vocabulary. With the lack of congressional approval, non-essential federal institutions and programs were either shut down or faced temporary non-essential personnel dismissals. Hiring freezes were again employed to reduce expenses. Not only was the economic impact severe, but the daily operations of military installations were restricted.

Another round of Army personnel cutbacks affected all Army installations. Even Fort Leonard Wood was faced with personnel losses on the base. On February 20, 2014, Major General Leslie Smith, Maneuver Support Center of Excellence and Fort Leonard Wood commanding general, announced the loss of approximately 1,000 Forces Command Soldiers and about 180 Department of the Army civilians. This was the result of the 2020 Forces Structure Realignment Study recommendations.

Base Activities

In January 2013, Secretary of the Army John McHugh and Chief of Staff of the Army General Ray Odierno announced that army base operations were to be reduced by 30%, adding, "We expect commanders and supervisors at all levels to implement both the guidance contained in this memorandum and the detailed instructions to follow. The fiscal situation and outlook are serious." Later that same year another government shut-down occurred with many base personnel forced to take an extended leave of absence. These actions had a great impact on the daily operations of Fort Leonard Wood, again with limited funding and personnel.

In spite of these challenges, Fort Leonard Wood continued in its role of innovation and meeting calls for service. 2010 ended with a natural disaster on the base. On December 31, a devastating EF3 tornado hit the base's housing area, destroying forty houses. Fortunately, because of the holidays, the base housing was nearly empty. About one year later, a ribbon-cutting ceremony was held for the replacement of the homes in the Woodland New Homes Expansion Project. The entire project with a total of 160 duplex houses was completed in May 2013.

That same summer, soldiers from Fort Leonard Wood also volunteered to assist with local flooding relief efforts in the communities surrounding the base. In December of 2011, soldiers from the base answered the Army Corps of Engineers call for help in the aftermath of Hurricane Sandy on the U.S. East coast.

Fort Leonard Wood continues its physical growth in the 2010 decade. The first Geospatial class began in January 2012. In September 2013, a groundbreaking ceremony was held for the construction of a new National Guard training facility for the 140th Regiment, the Missouri Regional Training Institute. The new $18 million, 53,251 square-foot facility will house an auditorium, and a multipurpose room administrative and support areas. In August 2014, ground was broken for a new $85 million dollar Advanced Instruction Training Center, to be completed in 2016.

Continued refinement of the military advancement programs was assisted in early 2014 with the completion of the link to "the Structured Self Development, of SDS training, with professional military education courses for promotion eligibility."

Among the training centers on base are the U.S. Army Engineer School; the U.S. Army Chemical, Biological, Radiological, and Nuclear (CBRN) School; and the U.S. Military Police School. Also added as a specialized area of instruction is the transportation training facility. Each held a significant purpose and worked in a coordinated effort to make the most efficient use of manpower and facilities.

Engineer School

The engineer school is the oldest of the training centers on base. Its purpose was defined to "provide freedom of tactical, operational and strategic maneuver to our forces and deny it to our enemies." It has been further described as enhancing the lethality of artillery fire, and protecting our combat power from the enemy and the environment while employing "Apollo 13 engineering" everyday. It is also seen as "the center level executive agent for engineering operations, defeat the device in C-IED, environmental integration, base camp development and operation, and geospatial engineering." It is the focal point for brigade engineer battalion implementation.

The training levels in the early 2010s included some 2,660 students daily in 80 courses. Annually over 1,200 are trained for deployment in C-IED operations at its explosive hazards center. An additional

300 combat engineers are trained annually at its demanding Sapper leadership course.

Chemical/Biological School

This school of training has become increasingly more important in the 2010s. This specialized training includes privates through colonels in an interservice center, providing education for "first responders, civil support, all hazards assessment and response, combating weapons of mass destruction." It is also the only U.S. military facility for toxic agent training.

Military Police School

As an interservice center, training is provided for the military and international partners and is a fully accredited law enforcement training institution. The expanding role of military police beyond security and policing, requires specialized training in working with nationals and reconnaissance.

Transport Training School

The newest of the national interservice training centers offers training for operators and maintenance of vehicular transport machines for the military. Heavy-duty equipment, such as earthmovers, require specialized mechanical knowledge for needed maintenance in theaters of combat. All military transports (trucks, tanks, trailers, etc.) require routine and emergency maintenance needs.

Fort Leonard Wood Today

The evolving growth of Fort Leonard Wood presents a picture of a premier Army center that trains more than 85,000 military and civilians each year. In its continuing role of providing basic training, Fort Leonard Wood now trains 20% new soldiers and 25-30% of women in the Army. It is one of two gender-integrated reception sites and home to the Army's largest, most diverse Non Commissioned Officers Academy.

Fort Leonard Wood conducts discretionary training sessions of interservice and joint training missions for the U.S. Marines, U.S. Navy, and the U.S. Air Force. All Department of Defense (DoD) training for earthmoving, truck driving, civilian support and CBRN (Chemical,

Biological, Radiological, and Nuclear) first responders happens at Fort Leonard Wood. It is also the home of the DoD Humanitarian Demining (ordinance) Training Center with a large international student involvement, annually training 450 students from 70-80 countries. In addition, about 12-13,000 reservists from 120 reserve units are trained on base annually.

Recent Forbe's list of the fastest growing small towns in America ranked Fort Leonard Wood as third. For the past several years more than 60% of military retirees from Fort Leonard Wood chose to reside in areas around the base.

Fort Leonard Wood has an operating budget of $470 million in addition to the $680 million in annual military salaries. Continued construction projects on base total more than $262.7 million. The direct economic impact of the military base in the surrounding areas is about three billion dollars annually.

Fort Leonard Wood supports a very large population that currently includes the following:

- 6,700 military permanent party
- 12,700 military family members
- 12,000 military in training
- 7,600 civilian workers
- 58,000 retired military and their families

To meet the needs of this population, the base has established a campaign plan with goals to "support current operations, develop leaders, train technically and tactically proficient service members and civilians, design/develop and integrate future forces, capabilities and concepts; and take care of people and ensure quality of life."

Great efforts are made to involve the community and visitors regarding the history and significance of Fort Leonard Wood. For example, well-organized base museums offer excellent perspectives on three of the main training centers—their history and purposes. Some original barracks have been maintained and available for public inspections. The Public Affairs Office on base offers the opportunity for tours of the base and information requests. Numerous annual base-sponsored activities welcome participation by the community and the retired veterans. Families are consistently present, and encouraged to attend, regular graduation ceremonies.

Accomplishments

In April 2014, Missouri U.S. Senator Claire McCaskill, the highest ranking woman in the history of the Senate Armed Services Committee, described Fort Leonard Wood's mission as critical to the nation's military and future. She further stated, "Whether it's hazardous waste... IEDs... corrections... military police, whether it's all engineering, this is a base that is crucial to our nation's strength. Fort Leonard Wood is where a lot of the most important technical training that is forward looking occurs."

One of the real evidences of the accomplishments on base is found in the current leadership. In the beginning decades of the installation's history, there was discrimination due to race and sex. As of this writing, the base's commanding general is African American, and the highest-ranking civilian leader on base is a woman. Major General Leslie C. Smith is a very personable and respected military leader. Dr. Rebecca Johnson has been at the base for twenty-two years and risen to the position of deputy to the commanding general. She is very active as the point of continuity for base operations and instrumental in base planning for the future.

Conclusion

Although many Missourians know *of* Fort Leonard Wood, few Missourians know *about* Fort Leonard Wood. The strategic importance of this military installation requires the knowledge and support of the local communities and the citizens of our State. That support is most effective when citizens become aware of the base's history and significance.

Chapter Eleven
FUTURE

Introduction

The history of Fort Leonard Wood presents an amazing record of purpose and commitment on the national and local level. As the purpose and commitment of military personnel and resources change on a national level, it will be seen on local levels in significant ways. Leadership, facilities and training techniques will be challenged to meet our evolving national military needs.

At the end of 2013, Major General Leslie Smith, who assumed command of Fort Leonard Wood in June, noted, "Now, with the U.S. withdrawal from the Middle East, recruits are experiencing a more traditional 'expeditionary-based' training program… They don't drive in and stay in a hut at a mock forward operating base, they stay in a tent." These short-term deployments reflect a newer and more relevant provision of military service.

World and National Views

The United States has become a war weary country. Involvement in Iraq and Afghanistan has been costly in terms of American lives and finances. The long-term results of combat operations are still being evaluated, but appear to challenge the initial reasons for our involvement. The United States military is becoming more defined and identified in its advisory and weaponry support roles.

Changes in the military are constantly being considered and evaluated. There has been much criticism of military structure as it exists today:

> Byzantine bureaucracies compromising dozens of overlapping command structures stifle innovation, slow response time, and create needless barriers.

Recruiting and retention processes designed for the 1970s frustrate many military personnel who expect a 21st-century employer.

Further frustrations have been expressed towards the "stop-loss" policy that invalidates contractual agreements and the "brass-creep" that has resulted in a top-heavy army. With the critical observations of the inherent challenges facing the military, there are also encouraging words of "future command structures and reorganization of personnel by skill set."

General Raymond Odierno, the Army's chief of staff, is a voice for such a change. He is leading the discussion for initiating the change toward "regionally aligned forces (RAF), with these regional forces having extensive training in language and culture and will establish long-term relationships with particular combatant commands."

But such ideas, as expected, receive a wide range of reactions from the military leaders. As one military leader responded:

> Everyone I speak to seems to have a different understanding of what RAF is. Some see it as a long-overdue transformation of the whole army into an agile, culturally sophisticated force; others see it as a tool of imperialistic intervention. Still others see it, in the words of one former Pentagon official, as "another giant army nothing-burger."

It is certain that with international, economic and political pressures, the U.S. military will change its structure and operations. With withdrawals from Iraq and Afghanistan, and politicians desperate to cut costs, the U.S. Army "as a whole is struggling to define—and defend—its role and mission." It is reassuring that U.S. military bases, like Fort Leonard Wood, have adapted to change in the past and will be able to lead in the changes in the future.

Technological Advances

Advances will continue to be made in weaponry and security for us, our allies and our enemies. While the U.S. continues in its development of more efficient military technology, that information

Future

will be shared with our international allies. More than likely, through cyber-espionage, our enemies may also be privy to these new discoveries.

It is unlikely that the United States will become involved in a major conventional conflict. It is more likely that the "American military will more probably continue to undertake peacekeeping efforts and perform humanitarian operations on relatively large scales of division-size or greater commitments."

Weaponry

Continued refinement of computer-enhanced weapons, such as armed drones, smart bombs and missiles, will require an efficiently trained soldier. New weapons technology will require new advanced training courses. The modern soldier will have more financial investment made in his or her training. The soldier with this specialized training and knowledge will need to be compensated to ensure a longer service to the military.

Security

The primary responsibility of national defense will require a continued, aggressive watchful eye to proactively counter acts of terrorism rather than simply responding to acts of enemy aggression. The security of the American people and property is the foremost purpose of the U.S. military and the highest priority in budgeting and operations.

Beyond the potential physical attacks on civilian personnel and property, there must be a strong commitment to cyber security. It is noted that, "Cyber-warfare and cyber-terrorism may represent the most dangerous threat to American national security in the years to come." Use of computer-assisted weaponry, communications and video observations will require increased attention to the security of computer software and programs.

Although the number of troops on the ground in future combat situations may decrease in size, the safety of those combat troops will continue to be an ongoing concern. New protective clothing and equipment will be needed to meet the new local enemy threats.

As the number of U.S. military bases decline in number and increase in importance, continued new plans for their physical protection will need to be in place. Recent rare incidents of acts of violence on our military bases indicate the continued need for screening of personnel and scrutiny of potential personality problems.

Fort Leonard Wood and the Future

Just as the U.S. military establishment will continue to adapt to new methods and procedures for war, so Fort Leonard Wood will also continue to adapt its training and operations to meet those needs. Hopefully, the threats can be answered before they materialize. However, the realities of national concerns over military budgets and operating expenses will continue to pose challenges for bases such as Fort Leonard Wood. It was reported recently that, "The Army estimates the southern Missouri installation could lose more than 5,000 civilian and military positions by 2020 after a round of much smaller cuts last year. The Army anticipates trimming its overall forces by more than 140,000 in that time period after 13 years of war in Iraq and Afghanistan."

The future of Fort Leonard Wood may seem secure in light of its ever-increasing significance on a national level. But that significance does not exempt it from the continued decreases in operating budgets. It will require a continued strong voice of support from the public and politicians.

Fort Leonard Wood is also being very proactive in evaluating and planning for social, environmental and economic sustainability. It is showing its traditional initiative in planning for strategies to develop sustainability and demonstrate its ability to become an economically viable asset for the U.S. military. There is a focus to create "net Zero" by 2025—to create as much energy as it uses and to return water cleaner than received. Programs being explored concerned efficient land use development, a zero electric truck and v-RIDE, securing "foodshed" to produce and recycle food on base, and use of building deconstruction materials. These programs will reduce the annual cost of operations and make the base a more attractive military installation regarding efficient funding.

Another important aspect regarding the future of Fort Leonard Wood lies in its ability to define its public image. In the past, organizations such as the Committee of Fifty have served as effective public relations for the installation in the community and on a national level. A new organization created in 2011 has assumed the current voice for the Fort Leonard Wood region. The Sustainable Ozarks

Partnership (SOP), which promotes economic development in the four counties that surround Fort Leonard Wood, is a strong focus on speaking about and for the base. In the spring of 2012, SOP held its first annual conferences. These conferences bring leaders from the military, community and Missouri politics to discuss means of better promoting the Fort Leonard Wood region. One example was encouraging the public to weigh in with comments to the Army leaders about the predicted cutbacks in personnel and funding. Over 4,400 responses were directed to the Army leadership. This organization is growing in its influence and reputation as an innovative leader in community and military promotion.

The public can offer their support through learning more about the Missouri base. That learning can come from reading, visiting and communicating with base personnel and veterans. The public can also make their voice heard through communications with local and state politicians about the importance of the military installation and the negative impacts of inadequate funding and personnel cuts.

Conclusion

Much of the direction for Fort Leonard Wood and other U.S. military bases will be determined by the nature of the threats and enemies that will be faced in the years ahead. Our country must be prepared to respond in an efficient and effective manner. As national and international cooperation in sharing intelligence findings improves, it is hoped the actions of the enemies of freedom may be stopped before major incidents occur.

Conclusion

Military Lessons

It has appeared that each decade in our national history has seen a change in military challenges with new definitions of combatants and combat procedures. The traditional warfare of WWII and Korea yielded to the experiences of cold war operations and then the jungle warfare of Vietnam. Following that came an evolutionary and ongoing approach dealing with terrorist activities having less nationalistic identities and more religious extremism. New combatants such as ISIS, and alShabbah followed AlQueda and Hamas. It is likely that newer and more radical groups will rise up to threaten peace and security around the world in the months and years ahead.

In the final analysis, some military operations were more successful than others. While some military leaders would decry the involvement of political control in combat operations, it is a reality and necessity in today's geo-political world. Some military operations had more of a sense of national support and sacrifice than others. A smaller percentage of the American public is now involved directly with military operations. The stresses of that minority of Americans affected directly by military combat operations also increased. It is noted that during operations in Iraq and Afghanistan over two million Americans did at least one tour of duty, with many serving multiple tours in one or both theaters. It was also noted that "Spending long periods in the combat zones and away from families put incredible stresses on the most patriotic and loyal American servicemen and servicewomen in the All-Volunteer Force." The resulting cases of post-traumatic stress syndrome and suicides are warnings of the dangers that are affecting veterans and also affecting recruitment.

The definition of victory in warfare is also being much more difficult to establish. While the United States certainly possesses superior weaponry with overwhelming force in combat, the failure to maintain the peace in the global stages of engagements was more common.

Conclusion

Base Contributions

Fort Leonard Wood made major contributions to every combat operation our country experienced since WWII. Besides the important basic training of the American soldier, there became a need for more specialized training in the development of new technology. With expansion on the base, came greater responsibility for the soldier and to the country. As the importance of the base's training centers expands, the impact reaches far beyond the state borders.

State and National Pride

It is interesting to consider what General Leonard Wood would think of today's national military situation in general and the Army base named after him. He likely would be very pleased at the initial response to the establishment of the base in Missouri, and the thousands of workers who built the installation under adverse weather conditions and in such a brief amount of time. He would be proud of a community that rallied to support the permanency of the base and he would be pleased with the continued role of civilian support services provided for in the base's daily operations. General Wood would be amazed at the growth of the base in gender-integrated and interservice expansion.

When examining the recent national military operations, General Wood would probably express some disappointments. He would most likely be surprised and disappointed at the small percentage of American soldiers and families directly involved in military combat actions. He would be disappointed in the general lack of knowledge that the public has about the installation bearing his name.

Considering General Wood's original concern for national military preparedness, he would be well-pleased with the base's operations today. He would also very likely be extremely proud of his name being associated with the mid-Missouri military training base. General Wood would also encourage the continuing military support of influential political leaders from our state. We should share in his concerns and his pride.

The Role of Civilian Support

Civilians have always had a strong voice in the role of the U.S. military. There was immediate support of troops in WWII as has been noted throughout this book. There was eventual negative reaction in the prolonged wars in Vietnam and Afghanistan. While military actions should never be determined by national polling, politicians and the military should be very conscious of the level of national support. Politicians are very consumed with the prospects of reelection that the voice of their constituents is very important on matters of national budget and military needs.

"Looking back in history, it was a defining moment in our Republic when after the Revolution, General George Washington took off his uniform and resigned his military commission. That example gave the message to future Americans that the military is subservient to civilian authority. That's a lesson that stands with us today."

(The Honorable Ike Skelton on Civil-Military Relations, November 2007)

With the primacy of civilian authority comes greater responsibility for that authority. Civilians must be more aware of the needs, actions and responses of our national military involvement. The political leaders must be more responsible in the deployment of our national military. When such deployments are deemed necessary, there must be national call to arms in support of our troops. There should be a more meaningful way to support out troops than being called on to "go shop in the mall." There should be a shared responsibility among civilian families instead of only a very small percentage of American families feeling the direct impact of war.

Ike Skelton also warned us in the following:

> Only one-half of one percent of Americans are in uniform today, although a good number of them are in the National Guard and Reserves and are being deployed frequently, which brings it home to many communities. I remember very, very well how with the general mobilization in the Second World War, in Korea, and in many respects Vietnam, the general population was touched more. In those days we all had been in the service or knew someone who was. Today, in many communities, military presence is an exception. I think that we as a country lose by that, especially if it makes us less conscious of the costs of freedom.

Conclusion

Civilians have also had an important role in the history of Fort Leonard Wood. At its creation 75 years ago, over 30,000 civilians joined together to build an Army base from the woods in winter weather in five months. After WWII, local civilians joined together to lobby effectively to make the base a permanent military installation. In the aftermath of the tragedy on September 11, 2001, local civilian workers rallied to work many extra hours to provide assistance in round-the-clock preparations on the base's combat operations centers. Today, as the Army considers cutbacks and realignments and closures, it will require a strong civilian voice to keep Fort Leonard Wood at its current level.

Very simply, just as the military is to defend the American public, the American public—especially Missourians—need to defend the military. We are interdependent as Americans.

Acknowledgments

I offer the greatest thanks to Dr. Larry Roberts, retired base historian for Fort Leonard Wood from 1989 through December 2011. He was gracious in his contributions of procedural advice, offering use of his printed historical materials and suggested contacts on and off base. Also very helpful was Dr. David Ulbrich, Engineer School Historian. He opened his office to me and gave needed guidance in negotiating on base procedures. Reserve Lieutenant Colonel Renea Timko, part-time base command historian, was also helpful in making me aware of research resources on and off the base.

Research at Fort Leonard Wood was made easier by the kind assistance and encouragement of the three aforementioned. I was also assisted in research by the staffs at the Truman Presidential Library, numerous Springfield, Missouri public libraries, and the Kansas City public library.

I am appreciative of all of the individuals who gave me their time and stories about experiences at Fort Leonard Wood and in the area. Peggy LaJuene was gracious in providing me access to wonderful family photos of Fort Leonard Wood. Ron Selfors and Keith Pritchard, with the Sustainable Ozark Partnership, were kind to offer me assistance in interviews with local citizens in the Waynesville/St. Robert area.

In the writing process, a number of friends and family have been helpful in proofreading. Among these were my wife, Jan, Gaytha and David Suits, Harold Holden, Reba Parker and Ann Willingham.

I am honored to work with *Missouri Life* Media. Editor-in-chief Danita Allen Wood and her husband, Publisher Greg Wood, have been gracious in their encouragement and promotion of this book. Doug Sikes and Randy Baumgardner were instrumental in editing, designing, publishing and promoting this book.

Resources

Books:

Austin, Roger O. *One Man's War*. Austin Books, 2009

Eisenhower, John S. D. *Teddy Roosevelt & Leonard Wood: Partners in Command*. University of Missouri Press: Columbia, Missouri, 2014

Fowle, Barry W. *Builders and Fighters: U. S. Army Engineers in World War II*. Office of History, United States Army Corps of Engineers, Fort Belvoir, Virginia, 1992

Hamby, Alonzo L. *Man of the People: A Life of Harry S. Truman*. Oxford University Press: New York, 1995

Johnson, Paul. *Churchill*. Viking Press: New York, 2009

Lane, Jack C. *Armed Progressive: General Leonard Wood*. University of Nebraska Press: Lincoln, Nebraska. 1978/2009

Mayes, Don. *Ft. Leonard Wood, Missouri, 1941*. U. S. Government Printing Office: 1991-654-002/43-461941. Reprinted for the 50th anniversary of Fort Leonard Wood. Clarke Library reference #255.7, Fort

McCallum, Jack. *Leonard Wood: Rough Rider, Surgeon, Architect of American Imperialism*. New York University Press: New York, 2006

Morris, Edmund. *Theodore Rex*. The Modern Library: New York, 2002

Muehlbauer, Matthew S. and Ulbrich, David J. *Ways of War*. Routledge, New York, 2014

Roberts, Larry D. US Army Engineer Center and Fort Leonard Wood Annual Historical Review (RCS CSHIS-6 R3) January 1, 1988 – December 31, 1998, two volumes

Skelton, Ike. *Achieve the Honorable*. State Historical Society, Columbia, Missouri, 2013

Smith, Stephen D. *A Historical Context Statement for a World War II Period Black Officers' Club: Building 2101, Fort Leonard Wood, Missouri*. Contract No. DACA88-97-M-0314, September. 1998

Smith, Steven D. *Made in Timber: A Settlement History of the Fort Leonard Wood Region.* ERDC/CERL Special Report 03-5, July 2003

Truman, Margaret. *Bess W. Truman.* MacMillan Publishing Company: New York, 1968

Wood, Eric Fisher. *Leonard Wood: Conservator of Americanism.* Forgotten Books: New York, 1920/ 2012

Newspaper Articles:

11/40	"Rolla Looks Like Army Camp"	*Rolla Herald*
11/7/40	"Army Surveys Site at Rolla For Camp"	*Rolla Herald*
11/14/40	"Army Reported to Have Decided…"	*Rolla Herald*
11/21/40	"Rolla Scene of Activities…"	*Rolla Herald*
12/40	"Army Camp Stimulates Telephone Service"	*Rolla Herald*
12/19/40	"Army Camp Doubles Newburg Responsibilities"	*Rolla Herald*
1941	"A 3-Week Contrast in Building…'	???
1941	"Grant's Grandson to Ft. Wood"	???
1941	"Fort Wood's Out of the Mud"	???
2/1/41	"Mud Delays Ft. Wood"	???
2/7/41	"Ft. Woods By May 25"	???
2/9/41	"Building Fort Woods Brings…"	*Kansas City Star*
3/4/41	"Fort Leonard Wood, Missouri"	*The Military Engineer*
4/13/41	"FLW, Missouri's Great Army Training Center"	*Kansas City Star*
7/9/41	"New Ozark Cantonment To Be FLW"	*Springfield Daily News*
8/14/41	"Big Guns Fired on Fort Wood Range"	???
9/5/41	"Sally Rand, world famous fan dancer…"	*Pioneer News*
9/30/41	"General Warns of Bad Leadership"	???
12/6/41	"Lessons of US War Games…"	*St. Louis Post-Dispatch*
3/30-4/2/42	ERTC Extract	*US Army*
4/2/42	"Requirements and Replacement Rates…"	???
7/5/42	"What Soldiers Think"	*US Army*
8/6/42	"Misconduct of Military Personnel…"	*US Army*
8/6/43	"ERTC to Build New Bridge…"	???
	"110 Become Citizens in Mass Ceremony	???
9/1/43	St. Louis Cardinals play an all-star team"	???

11/12/43	"Colored Troops Enjoy Program"	???
10/18/44	"War Department Technical Manual	US Army
4/23/45	"Processing of Redeployment Personnel	???
1955	Sixth Armored Division Company Book	
1961	Fort Leonard Wood map	base phone book
7/8/66	"New Construction at FW"	Guidon
7/15/66	"FY 1966 Big Year for Post Growth"	Guidon
7/22/66	"Water Shortage Accompanies Heat"	Guidon
7/29/66	"'Year of the Test' at Wood"	Guidon
8/5/66	"Record Number of Inductees Arrive"	Guidon
8/12/66	"Quarry Machine Operator's Course"	Guidon
8/19/66	"Post Construction Halted"	Guidon
8/26/66	"US Soldier Serves in Many Ways (VN)"	Guidon
9/9/66	"FW Picketed Again"	Guidon
9/16/66	"No Civil Rights Problem at FLW"	Guidon
10/21/66	"Tornado Barely Misses FW"	Guidon
10/28/66	"FW Now Largest USATC"	Guidon
11/4/66	"Nation's Largest Training Center"	Guidon
12/16/66	"Jet Age Arrived at FW Sat."	Guidon
2/24/67	"25,000 Reservists Face Call-up"	Guidon
6/2/67	"Beaver's Brother Starts Basic Training"	Guidon
1/5/68	"Gen. Walker Orders Fort Woods Cutback"	Rolla Daily News
2/16/68	"Ft. Wood is 7th 'City' in State"	St. Louis Globe-Democrat
5/26/68	"Don't Move! It's booby trapped!"	Sunday News and Tribune
6/68	"Impact of Public Spending…FLW"	USDA
6/1/68	"Let Barracks Work"	Kansas City Times
10/25/68	"Unique Exercise Trains for VN…"	???
1/14/70	"Prospects Good, Despite 'Tight Money'…"	Fort Gateway Guide
1/21/70	"Concerted Efforts… Naming Post Permanent"	Hannibal Courier-Post
3/2/70	"Teacher Barred After Visiting FLW"	Independence Examiner
3/5/70	"Cuts Set for 371 Bases"	Kansas City Times
4/4/70	"Battle of FLW Involves Army, Politicians, …"	Kennett Missouri Weekly

8/27/72	"Fort Wood Welcomes First Black General"	St. Louis Post-Dispatch
9/1/72	"Clean Out Vipers' Nest"	St. Louis Globe-Democrat
9/7/72	"Pulaski County citizens wink at Crime—Danforth"	St. Louis Globe-Democrat
1974	"Strong Ties to Bloodland"	???
1975	"In the Beginning"	Guidon
1975	"Fort Wood Injects $15 Million a Month…"	Pulaski County Democrat
5/19/75	"$34 Million Contract Awarded for …"	Daily Guide
9/20/75	"DIs charged with blackmail, and bribery"	???
10/8/75	"Private charged with selling LSD"	Lebanon Daily Record
10/16/75	"Court Marshall for 9 sergeants"	Honolulu Advertiser
10/16/75	"Five Schools From Ft. Belvoir to Relocate…"	Pulaski County Democrat
12/17/75	"Ft. Wood recalled"	Lebanon Daily Record
11/15/76	"Rhinc Wine III Tests One Army Concept"	Daily Guide
10/6/77	"Walk into Ft. Wood's Past"	Guidon
1980s	"Fort Wood History"	Old Settlers Gazette
3/80	"But He Thought His Truculence…"	US Army
3/9/80	"Soldiers—until war do them part?"	Chicago Tribune
3/10/80	"WWII Buildings Are Removed From Fort"	Daily Guide
3/23/80	"Trainees march where trees grew"	Guidon
5/80	"Remember When"	Construction Adviser
7/80	"Civilian education in Today's Army"	all Volunteer
8/14/80	"Local Man helped build barracks"	Guidon
10/11/81	"Dru Pippin funeral"	???
12/17/81	"FLW Exodus"	Daily Guide
2/28/85	"PAO Release, Ft. Belvoir relocation"	FLW PAO
1986	"Introduction"	Base info book
12/11/86	"Army takes lead in reducing drinking, driving"	Guidon
12/11/86	"Education key to harassment problem"	Guidon
8/14/87	"Post to Break Ground"	Daily Guide
9/87	"Construction started for engineer school"	Employee News
11/25/87	"US Army contracts Irk Skelton"	Columbia Daily News
12/9/87	"Fort Wood named best TRADOC"	Jefferson City Post-Tribune
12/20/87	"Fort Wood Construction Through the Years"	Daily Guide

5/24/88	"Fort Wood is Army's best post"	*Lebanon Daily Record*
5/24/88	"Army plans cuts in personnel"	*Columbia Daily Tribune*
6/1/88	"Fort Wood welcomes engineer school"	*Lebanon Daily Record*
7/6/88	"Skelton: Engineering School…"	HoR news release
9/22/88	"Ceremonies rename post, main gate"	*Essayons*
9/29/88	"Tomorrow we are the Engineer Center"	*Essayons*
5/28/89	"Barracks"	*St. Louis Post-Dispatch Magazine*
5/28/89	"Not So Fond Recollections of FLW"	*St. Louis Post-Dispatch Magazine*
???	The good old days: Woman recalls…"	*Daily Guide*
3/25/90	"Grunts and Groans"	*St. Louis Post-Dispatch Magazine*
4/24/90	"A Special Dedication Day Edition"	*Essayons*
1991	"Engineering the Future"	Base info book
4/13/91	"Base closings to help FW"	*Springfield News Leader*
4/21/91	"Army picks investigator"	*Rolla Daily News*
4/28/91	"General explains FT contribution to DS"	*Lebanon Daily Record*
6/27/91	"1939 war fitters force US …:	*Essayons*
6/27/91	"POW"	*Essayons*
7/1/91	"Fort Leonard Wood and the Gulf War"	*Ft. Wood at 50*
10/16/91	"FW construction to reach $46 million"	*Springfield News-Leader*
12/4/91	"General cuts Huet-Vaughn sentence in half"	*Kansas City Star*
12/4/91	"Huet-Vaughn verdict stands, …"	*Springfield News-Leader*
4/7/92	"Anti-War Physician Released Early…"	*Springfield News-Leader*
11/23/92	"Book says Missouri ranked hardest hit…"	*Jeff City News & Tribune*
4/1/93	"Book says Missouri ranked hardest hit…"	
5/26/93	"FW Narrowly Avoids …Closure…"	*Fort Leonard Wood Constitution*
6/6/93	"The fort that would take McClellan"	*Anniston Star*
5/14/95	"Skelton Voices Support of School's Move"	*Daily Guide*
6/24/95	"FW to get chemical, MP schools"	*Jefferson City News & Tribune*

6/24/95	"Taps for Fort"	*Anniston Star*
9/21/95	"Get ready now for FW growth"	*Rolla Daily News*
10/10/95	"More Than $200 Million To Be Spent…"	*Lebanon Daily Record*
8/9/1998	"When The Ozarks Made Way For War"	*Ozark Mountaineer*
9/98	"Black Officers' Club"	*GPO*
9/23/99	"New era of training"	*Guidon*
9/27-10/1/99	"FLW Through Time" MSC, Inauguration Week	
11/12/99	"Army Life"	*AAA Midwest Traveler*
9/13/01	"Post responds to Tuesday's terrorism"	*Guidon*
9/20/01	"Commander calls for 'determination and focus,'"	*Guidon*
	"Praise for civilians"	
	"Nation calls on reserved during crisis"	
9/27/01	"Improper entry will bring 'stiff penalties'"	*Guidon*
	"No barriers, to cooperation by military, civilian, state"	
	"Airfields here tighten security"	
	"Computer security depends on each user"	
	"Off-post businesses pitch in"	
8/23/06	"Mo. Army base leader in gays discharged"	*Stars and Stripes*
11/16/06	"AAFES opens new main store:	*Guidon*
1/10/13	"A Year to remember"	*Guidon*
1/24/13	"Army Freezes hiring, cuts base ops"	*Guidon*
5/30/13	"Here Comes the Boom"	*Guidon*
7/31/14	"Reserve Marines on post for CBRN MOS training" *Guidon*	
8/28/14	"AIT Groundbreaking"	*Guidon*

Old Settlers Gazette Feature Articles:

7/21/83	"History of Fort Leonard Wood"	Pages 20, 25
	"Ties to Bloodland"	Page 27
7/26/86	"The Coming of Fort Leonard Wood"	Pages 20-23
7/25/87	"The Building of Fort Leonard Wood"	Pages 27-30
7/25/92	"Building the Railroad to Fort Leonard Wood"	Page 50
	"Fort Leonard Wood Postcards"	Pages 32-38
7/24-25/93	"Fort Hollywood"	Pages 23-27

	"A History of One Family's Service to the United States" Page 16	
7/30/94	"Fort Leonard Wood"	Page 42
7/25/94	"Fort Wood Boom Time"	Pages 42-44
7/31/99	"Fort Wood's Early Days"	Page 51
	"1948 Fort Wood Cattle Drive"	Pages 45-46
2002	"St. Louis Cardinals 3, Fort Wood Allstars 0"	Page 47
2005	"USO"	Pages 53-56
2010	"Roundup on Fort Leonard Wood"	Pages 44-46
2010	"Displacement of the Hill Folks"	Pages 32-43
2014	"Fort Leonard Wood, 1941"	Page 45

<u>Other Materials</u>:

Holmes, O. Wendell, "The Impact of Public Spending in a Low-Income Rural Area: a Case Study of Fort Leonard Wood, Mo." U. S. Department of Agriculture, Washington, D. C., June 1968.

Roberts, Larry D. Power Point presentation of the history of Fort Leonard Wood, Fort Leonard Wood Post Guides, 1969-1996.

<u>Personal Stories (Interviews/Correspondence/Writings)</u>:
- Roger Austin
- Dan Beach
- James Berry
- Leo Crabbs
- Doug Dollar
- Emery Elliott
- Fritz Ensslin
- Kenny Foster
- Dale Geise
- Clark Hale
- James Hromas
- Herman "Big Train" Jackson
- James Jameson
- David Johnson
- Virgil LaJuene

Paul Long
Virgie Mahan
Edsel Matthews
LeRoy Mellow
Bill Morgan
Albert Mussan
Bill Nebo
Jerry Obrey
Reba (Long) Parker
Al Parsons
Keith Pritchard
David Suits
James Wallace

Fort Leonard Wood Site Visits:
November 20, 2012
November 27, 2012
December 5, 2012
January 24, 2013
May 29, 2013
May 31, 2013
June 5, 2013
June 11, 2013
June 19, 2013
July 30, 2013
August 14, 2013
August 20, 2012
September 25, 2013
April 7, 2014
April 23-24, 2014
August 6, 2014
August 28, 2014

Other Site Visits:
Kansas City Public Library
Truman Library Archives, Independence, Missouri
Veteran's Home, Mt. Vernon, Missouri

Visitor's Center, St. Robert, Missouri
Westside Baptist Church, Waynesville, Missouri
Sustainable Ozarks Project 2015 Annual Meeting, Waynesville, Missouri
Old Stage Coach Stop, Waynesville, Missouri

Fort Leonard Wood Update

Since the book was completed, several major events of importance have occurred. First, in the summer of 2015, a new base commanding general was commissioned. Major General Kent D. Savre comes to the base with vast military experience both nationally and internationally. Second, Fort Leonard Wood continues its tradition of offering training for international personnel. About 450 international students are trained each year and within the past five years those students represented more than 100 countries. The base continues to be Missouri's fifth largest employer supporting 36,400 direct and indirect jobs, injecting $141 million into the local economy. There are now 2,355 buildings on post, providing 15.4 million square feet of training and support facilities.

Continuing ongoing, effective representation for Fort Leonard Wood in the surrounding counties and throughout the state, is the Sustainable Ozarks Partnership (SOP). In their December e-Newsletter, they listed their top 10 accomplishments for 2015:

- Wrote, approved and implemented a new Strategic Plan that identifies high priority strategies and actions that are considered best practices among installation/support organizations (Dec. 2014)
- Created and solidified a governance structure for SOP that ensures regional representation by a Board of Directors from the 4-county region
- Helped preserve basic training levels on FLW when they were threatened
- Worked to counter large proposed cuts under the Army's 2015 SPEA, rallied
- Over 1,200 attendees and state and Congressional leaders, prepared and delivered the "Listening Session" Presentation, and provided follow-on reports to the Army

- Helped secure the passage of state legislation creating a Military Advocate and appropriating additional money for a Washington, D. C. lobbyist
- Secured a seat on the Board of Directors of the Association of Defense Communities
- Helped Missouri achieve two additional policy issues on the USA4 Military Family initiative
- Achieved an agreement with Fort Leonard Wood to set up a Leadership Committee that will provide an ongoing partnership framework for mutual benefit of FLW, communities and state
- Provided advocacy for the Waynesville-St. Robert Regional Airport and engaged a study to determine possible strategies to ensure its growth and development
- Achieved sustainable funding for the year, especially through about $175,000 in regional business contributions

This remarkable organization continues not only to directly help Fort Leonard Wood, but is expanding its efforts to provide help and training for military veterans throughout the four counties.

Natural disasters in the state in the last days of 2015 and early days of 2016 have seen base personnel provide strategic assistance for relief organizations. Tragically, the natural disasters also claimed the lives of several of the international students on base.

The increasing efforts of on-base sustainability projects, continued effective actions by the Sustainable Ozarks Partnership and the growing awareness of the Defense Department leaders of Missouri's interests, combine to help create confidence in the future of Fort Leonard Wood. It is also hoped that this book will add to the knowledge and pride of Missourians to preserve and protect the military treasure we have in our state.

About the Author

Paul William Bass was born in Independence, Missouri. He graduated from William Chrisman High School, Southwest Baptist College and Midwestern Baptist Theological Seminary. He married Jan Smashey in 1969. He served in full-time church staff positions for nineteen years. In 1990-2007, he served at Ouachita Baptist University in Arkadelphia, Arkansas as director of student activities and held various adjunct teaching positions. He became the intercollegiate debate coach with OBU debaters winning three national championships. He took early retirement in 2007, and in that year received the Isocrates Distinguished Debate Coach Award from the International Public Debate Association. During that time in Arkansas, he served in several interim pastor positions and became a bivocational pastor at Anchor Baptist Church for twelve years.

He began his writing career in 2007 with publication of *No Little Dreams: Henry Garland Bennett, Educator and Statesman*. He received the Henry Bennett Distinguished Fellow Award from Oklahoma State University. In 2008 his second book in the Bennett trilogy, *Fellow Dreamers: Oklahoma, Education and the World* was published. The third book in the Bennett trilogy, *Touching the Dream: Point Four*, was published in 2009. That year he was commissioned by Oklahoma State University to write a history of their School of International Studies, entitled *Legacy, Leadership and Learning*. In 2011 he published two religious works from his pastoral experience, *In Jesus' Names* and *Minor Characters of the Bible*. In 2012 he published *Robert S. Kerr: Oklahoma's Pioneer King*. For the Kerr book, he received the 2013 Missouri Writer's Guild First Place President's Award and the 2013 Walter Williams Major Work Award. In 2013 he published *Grace through Tolerance*. In 2014 he completed the autobiographical work of *Me and Church*.

All of these works are available through NewForums.com and Amazon.com.

Paul and Jan are retired and living in Willard, Missouri. His email address is *bassp@obu.edu*.

Index

Numbers

1st Lt. Terry Facility 154
1st US Volunteer Regiment 65
2nd Engineer Company 120
3rd Battalion 120
3rd Brigade 167
4th Cavalry 35
4th Maneuver Enhancement Brigade 13, 153
4th Training Brigade 12, 124
5th Corps 38
5th Engineer Battalion 123, 169
6th Armored Division 11
6th Calvary 32
6th Infantry Division 50
8th Army Corps 33
8th Brigade 33
8th Division 50
10th Cavalry 32
10th Division 44, 45
15th Cavalry unit 33
43rd AC Battalion 125
43rd AG Battalion 128
49th QM Regiment 55
70th Division 50
72nd Field Artillery Battalion 49
72nd Field Artillery Brigade 50
75th Division 50
87th Engineers 171
89th Division 44
93rd Evacuation Hospital 169
97th Division 50
102nd United States Army Reserve Division 13
140th Regiment 189
515th Engineer Company (Pipeline) 169
560th Military Police Company 149
982nd Engineer Battalion 100
2020 Forces Structure Realignment Study 188

A

Aadland, Major General Anders 183
Aberdeen Proving Grounds 100, 180
Advanced Instruction Training Center 189
Afghanistan 182, 187, 193, 194, 196, 198
Afghanistan War 13
Africa 101, 104, 172
Albania 164
Alberici Construction Company, Inc. 128
Al Qaeda 13, 182, 198
alShabbah 198
Ambrose, James 161
American Expeditionary Forces 31, 33
American Legion 43
American Red Cross 112
Anniston, Alabama 173
Arizona 37
Arlington Cemetery 47
Armistice 45
Armored Combat Earthmover (ACE) 124
Army Corps of Engineers 24, 189
Army Engineering School 162
Army Engineers 19
Army Medical Department 35
Army Reserves 11, 12, 125, 166, 169, 178
Arthur H. Neumann Brothers, Inc. 24
Ashcroft, John 162, 174
Aspen, Lew 176
Austin, Roger O. 59
Austria 53

B

Badshaw, Walt 107
Bagnell Dam 58
Bain News Service 65
Ballard, Brigadier General Joe E. 170

Barracks, Building 1316 150, 151
Barracks Construction 84, 86
Barton, William E. 53
Base Air Service 96
Base Construction 12, 67, 68, 69, 70, 71, 72, 73, 74, 75, 76, 78, 79, 80, 81, 82, 83, 84, 87, 88, 94, 96
Base Cooks 95
Base Dance 89, 90, 91
Base Headquarters Complex 144, 145
Base Inactivation 62
Base Postal Service 92
Base Realignment and Closure 13, 173, 175
Base Tornado 13, 188
Basic Combat Training (BCT) 166
Bass, Paul William 214
Beach, Dan 123
Beardslee, Colonel 127
Belgium 17
Bellamine, St. Robert 100
Berlin 104
Berry, Alabama 24
Berry, James 24
Big Piney River 20
Bin Laden, Osama 187
Blaisdell, Major General Judd 177
Bloodland, Missouri 22, 23
Blossom Point, Maryland 180
Bolivia 122
Bosnia 164, 170
Boston City Hospital 35
Bracken, Eddie 53
Bradley, General 119
Britain 17, 53
Brodie, Maj. Alexander Oswald 65
Bryant, Anita 111
Bryan, William Jennings 32
Burke-Wadsworth Bill 18
Bush, President George 122, 173

C

Cambodia 164
Camp A. P. Hill 180
Camp Crowder 28
Camp Funston 44
Camp Robinson 18
Camp Vicars 33
Canada 32, 53, 111, 124
Capehart Housing Project 102
Cardini 53
Carnahan, A.S.J. 102
Carnahan, Governor Mel 102
Carter, June 111
Carter, President Jimmy 121, 122
Cash, Johnny 111
Casualty Assistance Command 176
Central African Republic 164
Central America 101
Central Missouri State University 117
C. F. Lytle Company 24
Chad 122, 170, 187
Chandler, Chick 53
Chaplin, W.W. 53
Chemical Biological Radiological and Nuclear (CBRN) School 13, 156, 180, 185, 189, 190
Cheney, Dick 173
Cheyenne, Wyoming 33
China 17, 40, 105
Chinese Boxer Rebellion 40
Christman, Major General Daniel W. 171
Civilian Conservation Corps 22
Civil War 22
Clarke Library 159
Cleveland, President 37
Clinton, President 175
Coast Artillery Division, The 41
Coast Guard 111
Cold War, The 11, 12, 97, 103, 105, 122
Cold War Training 101
Columbia 122
Columbia, Missouri 98, 107, 119
Combat Excavator (CEX) 124
Combat Gap Crosser (CGC) 124
Combat Mobility Vehicle (CMV) 124
Commissary 12, 117
Condit-Smith, Louise 36
Congo 104, 120, 164
Conreid, Hans 111
Cookville, Missouri 22
Coolidge, President Calvin 46, 47
Cooper, Helen 53
Corps Service Command 50
Cote d'Ivoire 183
Countermine Test Facility 166

Countermine Training Support Center 170
Counterobstacle Vehicle (COV) 124
Crabbs, Lt. Col. Leo B. 48
Cronkite, Walter 110
Cuba 33, 37, 39, 101, 104, 105, 170
Cuban Missile Crisis 12
Cyber-Terrorism 195
Cyber-Warfare 195
Czechoslovakia 53
Czech Republic 97

D

Danforth, John C. 119
Debo, Bill 113
Declaration of War 50
Department of Agriculture 20
Department of Army 125, 126
Department of Defense 52, 112, 162, 173, 184, 190
Department of Defense Humanitarian Demining Training Center 184
Des Moines, Iowa 24, 52
Detroit, Michigan 107
Diem, South Vietnamese President 105
Diller, Phyllis 111
Division of Customs and Insular Affairs 32
Division of the Pacific 36
Djibouti 183
Dominican Republic 104
Dow, Tony 111
Drury Center 123
Duvall Maintenance Complex 12, 125

E

Eagleton, Senator Thomas 116
East Coast Department 44
East Timor 164, 182
Ecuador 170
Eglin Air Force Base 180
Egypt 170, 187
Eisenhower, Dwight 105
Elihu Root 33
Elliott, Emery 9
El Salvador 122
Emergency Operations Center 183
Emperor William II 39

Energy Monitoring and Control System 169
Engineer Center 124, 125, 162, 170
Engineer Center Transition Office 127
Engineering, Bio/Chemical and Military Police Museums 146
Engineer Replacement Training Center (ERTC) 11, 21, 50
Engineer School 12, 124, 160, 178, 185, 189
Ensslin, Fritz 56, 57
Eritrea 183
Esquire 116
Essayons 125
Ethiopia 170, 183
Europe 11, 17, 29, 31, 34, 43, 44, 101, 170
Evening Shade, Missouri 22

F

Fellow Dreamers: Oklahoma, Education and the World 215
Field, Supreme Court Justice Stephen 36
Flowers, Major General Robert B. 176, 180
Ford, President 115
Foreign Liaison Office 124
Forney Field Airport 147
Fort A.P. Hill 126
Fort Assiniboine 32
Fort Bayard, New Mexico 32
Fort Belvoir, Virginia 12, 101, 110, 125, 126, 127, 128, 162
Fort Benning, Georgia 161
Fort Bliss, Texas 33, 161
Fort Dix, New Jersey 161
Fort Gateway Guide 120
Fort Gordon 179
Fort Huachuca 35
Fort Jackson, South Carolina 161, 179
Fort Knox, Kentucky 50, 161
Fort Lee 179
Fort Leonard Wood 26, 27, 28, 29, 30, 48, 49, 50, 51, 53, 54, 56, 58, 59, 60, 61, 62, 86, 88, 63, 64, 98, 99, 100, 101, 102, 103, 104, 106, 107, 108, 109, 110, 111, 112, 113, 115, 116, 118, 119, 120, 121, 122, 123,

124, 125, 126, 143, 127, 161, 162, 164, 165, 166, 167, 168, 169, 173, 174, 175, 176, 178, 179, 180, 183, 184, 185, 187, 188, 189, 190, 192, 193, 196, 202, 212, 213
Fort McClellan 173, 176
Fort McKinley 33
Fort McPherson, Georgia 36
Fort Polk 116
Fort Riley Military Reservation 44
Fort Rucker 179
Forts Eustis 179
Fort Sill, Oklahoma 106, 161, 167
Fort Stanton 32
Fort Whipple 35
Fort Wood News, The 52, 53, 62, 125
Fort Wood Sentinel 125
Foster, Kenny 124
Fox Amusement Corporation 25
France 17, 18, 124
Functional Area Assessment (FFA) 179

G

Gabon 164
Garrison, Lindley 42, 43
Gasconade River 26
Geise, Dale L. 99
Gender Integration 12, 167
General Leonard Wood Army Community Hospital 155, 175
General Sherman M4 US Army tank 152
Georgia 183
Georgia Tech 36
Geospatial Class 189
German POWs 56, 57
Germany 17, 18, 33, 40, 42, 43, 53, 106, 124, 128
Geronimo 187
G.I. Bill of Rights 62
Grace through Tolerance 215
Grant Hall 125, 154
Great Britain 17
Great Piney River 60
Greece 97
Greer, Brigadier General Edward 117
Grenada 122
Guam 33

Guidon 125
Gulf of Tonkin 106
Gulf Region 182

H

Haiti 164, 183
Hale, Clark 106
Hamas 198
Harding, Warren G. 46
Harney, David 171
Harris & Ewing 66
Hartzog, General William 177
Heflin, Senator Howell 174
Hitler, Adolph 17
Hoge, General William 128
Hoge Hall 158, 128
Holden, Harold 202
Homeland Security 183
Homestead Act of 1841, The 21
Honduras 122, 170
Hope, Bob 111
Horydczak, Theodor 66
House Armed Services Committee 116, 174, 175
Houston, Missouri 9
Hromas, James 112
Huet-Vaughn, Yolanda 171
Hungary 53
Hurricane Sandy 189

I

Illinois 176
In Jesus' Names 215
Indian Territory 37
Installation Management Agency (IMA) 13, 184
Installation of Excellence award 161
International Crises 164
Interservice Training Review Organization (ITRO) 12, 165, 167, 168, 180
Iowa 20
Iraq 12, 178, 182, 187, 193, 196, 198
Iraq War 12, 13
ISIS 198
Italy 17, 53, 170

J

Jackson, Herman "Big Train" 108, 109

Jaeger, James 100
JANUS 178
Japan 17
Jarboe Commission Company 63
Jefferson City, Missouri 59, 127
Jefferson City Post Tribune 118
Jerry Obrey 98
John B. Mahaffey Museum Complex 146
Johnson, David 113
Johnson, Dr. Rebecca 192
Johnson, President Lyndon B. 105, 106, 110
Joint Working Group 180
Jordan 170, 187
J.S. Alberici Construction Company, Inc. 128

K

Kansas City, Missouri 52
Kansas National Guard 107
Kawamura Hall 128
Kennedy, John F. 105
Kennedy, Robert F. 111
Kennett Missouri Weekly 116
Kent State 115
Kenya 164
Khanh, General Nguyen 105
King, Dr. Martin Luther, Jr. 110
Kirk, Phyllis 111
Kirksville, Missouri 19, 32
Kissinger, Henry 115
Kling, S. Lee 175
Korea 97, 99, 101, 103, 170
Korean War 11, 97
Kuwait 12, 166, 170

L

Laclede, Missouri 32
LaJuene, Peggy 202
LaJuene, Virgil 26
Lamour, Dorothy 111
Laquey, Missouri 117
Latin America 104, 121
Laughlin, Bill 64
Lebanon 120, 122, 183
Lebanon, Missouri 53, 63, 109
Lee, Henry 43
Legacy, Leadership and Learning 215

Leon, Iowa 19
Liberia 164, 183
Libya 187
Lieber Heights 146
Life Cycle Replacement program 169
Lincoln Hall 159, 160
Little Rock, Arkansas 18
Lockheed F-117 Nighthawk 177
Lodge, Congressman Henry Cabot 37
Lollobrigida, Gina 111
Long, Paul 51
Louis, Joe 53
Lusitania 43

M

MacDonald, Taylor 65
Macedonia 164
Made in the Timber: A Settlement History of the Fort Leonard Wood Region 21
Mahaffey Museum Complex 146, 149
Mahan, Virgie 172
Main Post Chapel 153
Mali 187
Maneuver Enhancement Brigade 184
Maneuver Support Battle Lab 178
Maneuver Support Center (MANSCEN) 13, 176, 180, 183
Maneuver Support Center of Excellence 13, 184, 188
Manhattan, Kansas 107
Marine Corps 161, 165, 168, 179
Marine Headquarters building (Marines of the Ozarks) 160
Mark Twain National Forest 20, 21, 22
Marshall, George C. 18, 58
Maryland 126
Matthews, Edsel 108, 109
Mauritania 170
McCaskill, Senator Claire 192
McHugh, John 188
McKinley, President 37
Me and Church 215
Medal of Honor 36
Mellows, LeRoy 120
Mexico 33, 42, 53
Michigan 176
Miles, General Nelson A. 36
Military Affairs Committee 28

Military Police (MP) School 13, 156, 180, 185, 190
Military Policy of the United States, The 41
Military Working Dog Section 170
Militia Affairs Division 41
Miller, Joe 100
Million Dollar Hole 123
Mine Detection Facility Working Group 179
Mine Dispensing Vehicle (MDV) 124
Minnesota 19, 49, 107
Minor Characters of the Bible 215
Missouri 17, 48, 127
Missouri Life 202
Missouri Ozarks 17
Missouri Regional Training Institute 189
Missouri Special Olympics 124, 172
Mitchell Glass Company 113
Mobile Affairs Division, The 41
Moellering, General John 126
Montana 32
Morgan, Bill 113
Morgan, Bob 63
Morgan, Dormalee 63
Morgan Sisters Singers 53
Morrisey, Chris 175
Mussan, Albert 23
MWR Outdoor Adventure Center 145
My Experiences in the World War 34

N
National Defense Act 43
National Defense Council 19
National Guard 11, 12, 43, 47, 107, 125, 161, 166, 169, 178, 189, 200
National Security League 42
Naval Air Warfare Center 180
Nazi Military Aggression in Europe 11
NCO Academy 183
Netherlands 17
Newburg, Missouri 21, 23, 25, 27, 29, 77, 63, 100
New Mexico 37
New York 42
New York City, New York 43, 182
New York Times 60

Nicaragua 170
Nigeria 182
Ninth Armored Division 128
Nixon, President Richard M. 111, 115
No Little Dreams: Henry Garland Bennett, Educator and Statesman 215
Non Commissioned Officers Academy 190
Noriega, General 122
North Korea 97
North Vietnam 106
Nutter Fieldhouse 147

O
Odierno, General Raymond 188, 194
Ofederstrom, Major A.F. 93
Officers Club 12, 117, 169
Ohio 46
Oklahoma 37
Oklahoma City, Oklahoma 33
O'Malley, Frank 120
O'Neill, Representative Thomas (Tip) 116
Operation Bright Star 170
Operation Desert Shield 12, 164, 169, 170, 178, 179
Operation Desert Storm 12, 164, 169, 170, 175, 178, 179
Operation Enduring Freedom 13, 182
Operation Iraqi Freedom 13, 182, 187
Operation Neptune Spear 187
Operation New Dawn 187
Operation Promote Liberty 170
Operation Restore Hope 170
Operation "Rolling Thunder" 106
Operations Other Than War (OOTW) 166
Owens, Buck 111

P
Pakistan 187
Palace, Missouri 22
Panama 122, 170
Paris Peace Talks 111
Parker, Reba 99, 202
Parris, Representative Stanford 127
Patton, George S. 33
Pearl Harbor, Hawaii 51, 60
Pennsylvania 182

Pentagon 182, 183
Pentagon Papers 115
People's Republic of China 97
Pershing, General John J. 31, 32, 33, 34, 40, 44, 47, 54, 66
Persian Gulf 169, 170
Peru 122
Philippine Division 39, 40
Philippines 33, 46, 122, 183
Pippin, Dru 11, 102
Plattsburg Barracks 42
Plattsburg Movement 34
Poland 53
Portugal 17
Post Exchange 12, 117, 169, 184
Prairie Mound, Missouri 32
Pritchard, Keith 120, 202
Puerto Rico 33, 170
Pulaski, Count Casimar 22
Pulaski County, Missouri 21, 23, 26, 63

Q
Quartermaster Detachment 55

R
Railway Construction 77
Reagan, President 122
Regimental Training Areas 11, 103
Remagen Bridge 128
Reserve Component 12, 170, 178
Reserved Officers Training Corps (ROTC) 47
Reserves 47, 200
Reth, Colonel Thomas B. 162
Reynolds, Senator Bob 28
Rhine River 128
Riggs National Bank 65
Robenson, Colonel 49
Robert S. Kerr: Oklahoma's Pioneer King 215
Roberts, Dr. Larry 7, 202
Robinson, Jackie 54
Robinson, Joseph T. 18
Robotics-Technology Insertion Activity 178
Rolla Chamber of Commerce 25
Rolla, Missouri 21, 25, 29, 31, 53, 63, 109
Rolla School of Mines 100

Rolling Heath School 124
Roosevelt, Eleanor 55
Roosevelt, President Franklin 17, 51
Roosevelt, Theodore 33, 37, 38, 39, 40, 42, 45, 65
Root, Elihu 41
Rose, Brad 175
Roskelly, Cpl. 100
Roubidoux River 20
Rough Riders 38, 47, 65
Route 66 21, 25, 29, 51, 102
Rumania 53
Russia 17, 42
Russo-Japanese War 33, 40
Ryan, Shella 53

S
Saigon 115
San Antonio, Texas 37
San Francisco, California 36
San Juan Hill 38
Santiago de Cuba 38
Santiago Province 38
Saudi Arabia 169, 170
Savre, Major General Kent D. 212
School of Defense Simulation Internet 178
Schroeder, Major General Daniel 178
Selective Service Act 18, 43
Selfors, Ron 202
Senate Armed Services Committee 192
Senate Military Affairs Committee 33, 45
Senator Clyde Herring 20
September 11, 2001 attack 13
Serbia 164
Seventh Corps 19
Seventh Corps Area Training Center 20, 28
Seventh Corps Regular Army Area Training Center 21
Short, Dewey 102
Sierra Leone 164, 182
Sioux City, Iowa 24
Sixth Armored Division 99
Sixth Division 49
Sixth Engineers 49
Sixth Infantry Division 19

Skelton, Representative Ike 127, 128, 162, 173, 174, 175, 200
Smashey, Jan 214
Smith, Major General Leslie 188, 192, 193
Smith, Steven 21
Somalia 164, 170, 183, 187
South America 101, 104
South Dakota 32
Southern Department 44
South Korea 97, 98
South Vietnam 105, 106, 115
Southwest Asia 170
Southwest Division 33
Soviet Union 97, 104, 164
Spain 17, 37
Spanish-American War 9
Special Entertainment Units 52
Special Services 120
Specker Barracks 12, 117
Springfield Chamber of Commerce 20
Springfield, Missouri 19, 24, 29, 52, 109, 123
Sputnik 101
Stalin, Joseph 61
Star Base 156, 157
State Normal School 32
Stimson, Henry 18, 41, 42
St. Joseph, Missouri 120
St. Louis Globe Democrat 118
St. Louis, Missouri 29, 52, 98, 128
St. Louis Symphony 53, 111
St. Louis Union Station 59
St. Robert, Missouri 100
Structured Self Development 189
Suits, David 117, 202
Suits, Gaytha 117, 202
Sustainable Ozarks Partnership (SOP) 196, 202, 212, 213
Sweden 17
Switzerland 17
Symington, Representative James 116
Syria 187
Systematic Productivity Improvement Review 165

T

Taft, President William Howard 40, 41, 42

Taliban 182
Tampa, Florida 38
Tanzania 164
Texas 37
Thurman Hall 156
Timko, Lieutenant Colonel Renea 202
Tokyo, Japan 33
Tomahawk Cruise Missile 177
Topeka, Kansas 107
Top Hat Club 119
Touching the Dream: Point Four 215
Training and Doctrine Command (TRADOC) 12, 166, 179
Training at Fort Leonard Wood 130, 131, 132, 133, 134, 135, 136, 137, 138, 139, 140, 141
Transport Training School 190
Tribune, Missouri 22
Truman Doctrine 104
Truman Education Building 171
Truman Education Center 12, 117, 123, 155
Truman Presidential Library 202
Truman, Senator, Harry S 19, 23, 25, 26, 27, 28, 48
Tulsa, Oklahoma 63
Turkey 187
Tuscumbia, Alabama 174
Tuskegee Air Flight program 55

U

Uganda 187
UH-1B "Iroquois" Helicopter 149
Ukraine 53
Ulbrich, Dr. David 202
Underwood & Underwood 65
Union of Soviet Socialist Republic (USSR) 105
United Nations 97, 101, 104
United Nations Security Council 97
United States 17, 18, 26, 28, 33, 34, 39, 43, 44, 46, 48, 50, 56, 60, 61, 62, 97, 98, 103, 106, 119, 182, 184, 187, 193, 195, 198
United States Army Engineer Center and Fort Leonard Wood 161
United States Army Reserve Division 184

United States Army Training Center-
    Engineer 11
United States Volunteer Cavalry 37
University of Nebraska 32
Upton, Emory 41
Up with People 111
U.S. Air Force 105, 111, 161, 165, 167,
    179, 190
U.S. Army 17, 18, 34, 108, 111, 116,
    172, 176
U. S. Army Engineer Center 12
U.S. Army Engineer Museum 171
U.S. Army Engineer School (USAES)
    126, 189
U.S. Army Training and Doctrine
    Command (TRADOC) 117
USA Today 175
U.S. Cavalry 35
U.S. Marines 111, 190
U.S. Military Police School 189
U.S. Navy 111, 165, 168, 179, 190
USO (United Service Organization
    Inc.) 11, 52, 53, 55, 111
U.S.S. Maine 37

V

Vietnam 105, 107, 108, 110, 111, 112,
    120
Vietnam War 11, 12, 104, 111, 112,
    113, 115, 116
Villa, Pancho 33
Virginia 127

W

WAACs (Women's Army Auxiliary
    Corps) 52
W. A. Kilnger and Sons 24
Walker, General 112
Wallace, James 109
Wallace Swimming Pool 169
Walter Reed Army Institute 117
Walter Reed Hospital 34
Walton, Bud 120
War College Division 41
War Department 19, 21, 41, 44, 45, 56
War on Terrorism 12, 13, 164, 182, 187
War Productions Board 60
Warren, Helen Frances 33
Warren, Senator Francis E. 33

Washington, D.C. 31, 32, 33, 34, 36,
    44, 47, 65, 110, 111, 182
Washington, General George 200
Washington Post 116
Watergate 115
Wayne, "Mad Anthony" 21
Waynesville, Missouri 9, 20, 21, 25,
    29, 51, 52, 53, 63, 102, 103, 107,
    109, 120
Waynesville/St. Robert Chamber of
    Commerce 175
Waynesville-St. Robert Regional Air-
    port 213
Wentworth Military Academy 128
Western Contracting Corporation 24
West, Luther 171
West Point 32, 128
Wharton, Missouri 22
Wheeler, Gennerral Joseph 65
White Ripple Club 25
Williamsburg, Virginia 170
Williams Elementary School 117
Williams, Steve 107
Willingham, Ann 202
Wilson, President Woodrow 42, 43,
    44, 45, 65
Winchester, New Hampshire 35
Wisconsin 176
Wood, Danita Allen 202
Wood, General Leonard 19, 31, 34, 35,
    36, 37, 38, 39, 40, 41, 42, 43, 44,
    45, 47, 65, 66, 199
Wood, Greg 202
Woodland New Homes Expansion
    Project 188
Woodring, Harry 18
World Trade Center Towers 182
World War I 9, 17, 18, 22, 31, 54, 128
World War II 11, 34, 48, 55, 62, 64,
    97, 128

Y

Yemen 182, 183, 187
Yugoslavia 53

Z

Zaire 104, 120, 164
Zentner, Si 111